Petroleum Reservoir Engineering

Petroleum Reservoir Engineering

James Cameron

SYRAWOOD
PUBLISHING HOUSE

New York

Published by Syrawood Publishing House,
750 Third Avenue, 9th Floor,
New York, NY 10017, USA
www.syrawoodpublishinghouse.com

Petroleum Reservoir Engineering
James Cameron

International Standard Book Number: 978-1-68286-814-0 (Hardback)

Cataloging-in-Publication Data

Petroleum reservoir engineering / James Cameron.
    p. cm.
Includes bibliographical references and index.
ISBN 978-1-68286-814-0
1. Oil fields. 2. Petroleum engineering. 3. Oil reservoir engineering. I. Cameron, James.
TN870 .P48 2019
622.3382--dc23

TABLE OF CONTENTS

PREFACE

Petroleum engineering is a field of engineering that is concerned with the production of crude oil or natural gas. The areas of formation evaluation, reservoir simulation, reservoir engineering, drilling, etc. are crucial to petroleum engineering. Reservoir engineering is a branch of petroleum engineering. It strives to solve the drainage problems that arise during the production of oil and gas reservoirs in order to achieve a high economic recovery. Numerical reservoir modeling, well testing, drilling, PVT analysis of fluids, etc. are central to reservoir engineering. The specializations in reservoir engineering are surveillance engineering and simulation modeling. This book presents the complex subject of petroleum reservoir engineering in the most comprehensible and easy to understand language. It is a valuable compilation of topics, ranging from the basic to the most complex theories and principles in this field. It is a complete source of knowledge on the present status of this important field.

A foreword of all Chapters of the book is provided below:

Chapter 1- A petroleum reservoir is a subsurface pool of hydrocarbons that is contained within fractured or porous rock formations. Such reservoirs are classified broadly into conventional and unconventional reservoirs. This is an introductory chapter, which will discuss in brief about petroleum reservoirs and their engineering. It includes topics like petroleum traps, reservoir rocks, reservoir porosity and reservoir permeability, among others;

Chapter 2- An oil well predominantly produces crude oil and some natural gas dissolved in it. The hydrocarbons present in crude oil include alkanes, aromatic hydrocarbons and cycloalkanes, besides others. This unique mix of molecules determines the physical and chemical properties of petroleum, such as refractive index, viscosity, hydrogenation, paraffin wax content, etc. The topics elucidated in this chapter cover some of the important physical and chemical properties of crude oil;

Chapter 3- Millions of years of heat and pressure have led to the metamorphosis of microscopic plants and animals into hydrocarbons such as oil and natural gas. A trap is formed when buoyant forces that drive the upward migration of hydrocarbons fails to overcome the capillary forces of a sealing medium. Traps are classified into structural traps, stratigraphic traps and hydrodynamic traps. This chapter closely examines some of the crucial aspects of these different trap formations;

Chapter 4- Oil and gas reservoirs are subsurface pools of hydrocarbons that are contained in fractured or porous rock formations. These are alternatively called as petroleum reservoirs, and can be classified into conventional and unconventional reservoirs. An elaborate study of oil and gas reservoirs has been provided in this chapter, which includes topics that cover oil and gas traps, PVT analysis for oil reservoirs, gas condensate reservoirs, etc.;

Chapter 5- Enhanced oil recovery (EOR) refers to the implementation of techniques for ensuring the optimal extraction of crude oil from an oil field. Thermal injection, chemical injection and gas injection are the three primary techniques for EOR. This chapter discusses in extensive detail about the processes of enhanced oil recovery that includes primary, secondary and tertiary recovery, infill recovery, etc.;

Chapter 6- The study of the origin, accumulation, occurrence and exploration of hydrocarbon fuels is under the domain of petroleum geology. The evaluation of the varied elements of source, trap, reservoir, seal, maturation, timing and migration of sedimentary basins is integral to this field. The topics elaborated in this chapter include oil sands, oil shale, basin and petroleum system modeling, biomarker, etc. are vital for a complete understanding of petroleum geology;

Chapter 7- Petroleum exploration refers to the search for petroleum deposits beneath the surface of the Earth. It is under the domain of petroleum geology. Various methods are employed in petroleum exploration, such as geological survey, geochemical survey, remote sensing survey, drilling, etc. These have been extensively discussed in this chapter.

I would like to thank the entire editorial team who made sincere efforts for this book and my family who supported me in my efforts of working on this book. I take this opportunity to thank all those who have been a guiding force throughout my life.

James Cameron

Chapter 1

Petroleum Reservoirs and their Engineering

A petroleum reservoir is a subsurface pool of hydrocarbons that is contained within fractured or porous rock formations. Such reservoirs are classified broadly into conventional and unconventional reservoirs. This is an introductory chapter, which will discuss in brief about petroleum reservoirs and their engineering. It includes topics like petroleum traps, reservoir rocks, reservoir porosity and reservoir permeability, among others.

A petroleum reservoir or oil and gas reservoir is a subsurface pool of hydrocarbons contained in porous or fractured rock formations. Petroleum reservoirs are broadly classified as conventional and unconventional reservoirs. In case of conventional reservoirs, the naturally occurring hydrocarbons, such as crude oil or natural gas, are trapped by overlying rock formations with lower permeability. While in unconventional reservoirs the rocks have high porosity and low permeability which keeps the hydrocarbons trapped in place, therefore not requiring a cap rock. Reservoirs are found using hydrocarbon exploration methods.

Example of Sand Formation

An oil reservoir can be in a sand formation in which oil has accumulated. Zion's current exploration would be in a carbonate formation. Using sand as an example is due to most people's familiarity with sand. If you covered a sand beach with tons of sediments, you would have a potential sand reservoir. The only difference is that the loose sand grains would consolidate into a hard sandstone rock.

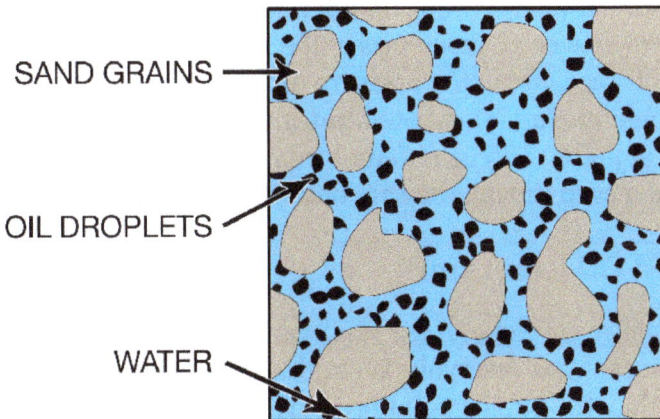

A microscopic view of oil bearing sand formation, showing how oil and water are found in between the individual sand grains.

Petroleum reservoirs are broadly divided into oil reservoirs and gas reservoirs. Oil reservoirs around the world are not the same; the fluid composition, the prevailing temperature and the pressure (called the reservoir pressure) all vary. We can further break down oil reservoirs into different types based on the interaction between this reservoir pressure and the hydrocarbon fluids.

Classification on the Basis of Phase Diagram

To understand this classification, we'll be looking at a very important chart; a pressure-temperature envelope or phase diagram. True, there are different types of crude oil, such as black oil, volatile oil, etc. For all oil reservoirs, the basic concept of the phase envelope is the same for each type of crude oil even though there are slight variations. Therefore, even though fluid types vary, we can study the three kinds of oil reservoirs by looking at a single phase envelope.

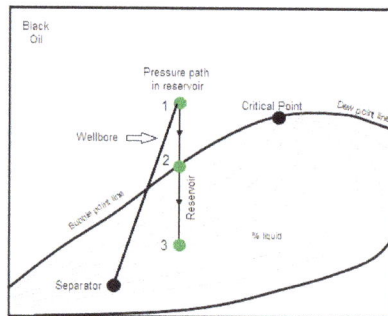

This is the phase envelope for an oil reservoir that contains black oil. The bubble point line on the envelope is our area of concentration for oil reservoirs. We already know that reservoir temperature and pressure for different oil reservoirs around the world varies. In this particular phase envelope, the numbers 1, 2 and 3 trace the different possible points the reservoir pressure can be. One reservoir in the North Sea can have the reservoir pressure at point 1 while another can have reservoir pressure at point 3. These two reservoirs may be oil reservoirs in the same region, but since they are found at different reservoir pressures, this pressure difference will affect the kind of oil reservoir and our depletion plan. Thus, our strategies to produce from a reservoir with a reservoir pressure at point 1 will be different from that of the reservoir whose reservoir pressure is at point 3 even though they are both oil reservoirs.

Undersaturated Oil Reservoirs

Take a look at the phase envelope. If the reservoir pressure is at point 1 then that reservoir is an undersaturated oil reservoir. What this means is that the reservoir pressure (point 1) is higher than the bubble point pressure.

The bubble point pressure (on the bubble point line) is a very important concept for every oil reservoir. This pressure is obtained at the laboratory from tests conducted on samples of the reservoir fluid. To conduct this test, a sample of the reservoir oil is placed in a chamber (container) at the reservoir pressure. We slowly lower the pressure exerted on this oil. It's not difficult to imagine, just like your soft drink or beer, there is always some gas bubbled into the liquid under pressure. This pressure is sealed by the bottle cap. When you open the bottle, you release the pressure and expose the liquid to the atmosphere thereby lowering the pressure, so the gases bubble and come out of the liquid. Let's assume that the bottling company did not force enough gases into this drink and the soft drink or beer could still have absorbed more gases if they were bubbled at that pressure. This assumption leads us to understanding undersaturated oil reservoirs.

The word "undersaturated" also means that this kind of oil reservoir is not saturated with gas bubbles. The saturation here is referring to gas bubbles, and so the pressure has kept all the lighter gaseous hydrocarbons (bubbles) inside the crude oil. At the same time, this undersaturated state tells us that assuming there were more gas bubbles, the crude oil in this reservoir would have comfortably absorbed them. An understanding of this will guide us in coming up with the best strategies to produce the hydrocarbons in this reservoir.

Saturated Oil Reservoirs

Any reservoir having a pressure that falls at point 2 on the phase envelope is a saturated oil reservoir. A saturated reservoir is fully saturated with lighter hydrocarbon gases at that reservoir pressure. Saturation means that the crude oil here is fully occupied with dissolved gases and will not take any extra gas bubbles at that pressure (point 2) unless we increase the pressure. Remember, it is the position of the reservoir pressure on the phase envelope that helps us distinguish what type of oil reservoir we are dealing with.

Back to our analogy with a bottle of beer or soft drink, in this case the bottling company kept forcing gas bubbles into the drink in the bottle until it could not take any extra gas. This state is the saturation state, and oil reservoirs found at this state are termed saturated oil reservoirs. It is not enough to know that we just drilled into an oil reservoir — knowing the kind of oil reservoir it is will help us choose the right methods to efficiently produce the hydrocarbons in the reservoir.

Gas-Cap Reservoirs

Point 3 on the phase envelope is called the 2-phase region; remember that at points 1 and 2 we only have crude oil in the reservoir with some form of gas saturation at that pressure.

In the beer and soft drink analogy, what happens the moment we open the bottle cap? Some gases come out of solution because we just disturbed the pressure equilibrium of the drink in the bottle. Gas-cap reservoirs are operating under similar conditions — some of the gases have come out of solution and are now occupying the top of the reservoir, whereas in the case of the soft drink or beer, the gases simply escape to occupy space in the room.

For an oil reservoir, because gas is less dense than oil, the gas will occupy the upper portions of the reservoir rock. If the reservoir pressure was higher (like at points 2 or 1) then the gases would have been forced back into solution. In other words, gas-cap oil reservoirs are supersaturated with gas. Gas-cap reservoirs have gases at the upper portions (cap) of the reservoir trap. Taking the time to know what kind of oil reservoir we are dealing with is never a waste of time, and this understanding could be the difference that allows us to maximize production from the reservoir.

Petroleum Reservoir Engineering

Petroleum reservoir engineering is the technology concerned with the prediction of the optimum economic recovery of oil or gas from hydrocarbon-bearing reservoirs. It is an eclectic technology requiring coordinated application of many disciplines: physics, chemistry, mathematics, geology, and chemical engineering. Originally, the role of reservoir engineering was exclusively that of counting oil and natural gas reserves. The reserves—the amount of oil or gas that can be economically recovered from the reservoir—are a measure of the wealth available to the owner and operator. It is also necessary to know the reserves in order to make proper decisions concerning the viability of downstream pipeline, refining, and marketing facilities that will rely on the production as feedstocks.

The scope of reservoir engineering has broadened to include the analysis of optimum ways for recovering oil and natural gas, and the study and implementation of enhanced recovery techniques for increasing the recovery above that which can be expected from the use of conventional technology.

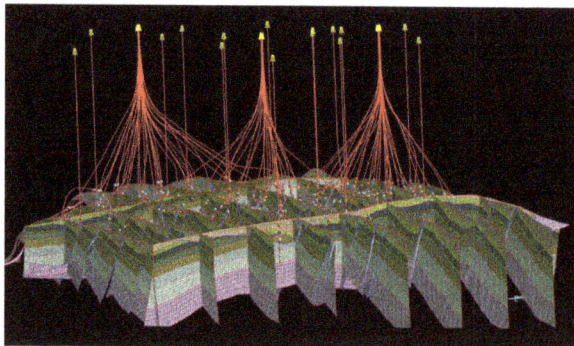

The amount of oil in a reservoir can be estimated volumetrically or by material balance techniques. A reservoir is sampled only at the points at which wells penetrate it. By using logging techniques and core analysis, the porosity and net feet of pay (oil-satu-rated interval) and the average oil saturation for the interval can be estimated in the immediate vicinity of the well. The oil-saturated interval observed at one location is not identical to that at another because of the inherent heterogeneity of a sedimentary lay-er. It is therefore necessary to use statistical averaging techniques in order to define the average oil content of the reservoir (usually expressed in barrels per net acre-foot) and the average net pay. The areal extent of the reservoir is inferred from the extrapolation of geology and fluid content as well as the drilling of dry holes beyond the productive limits of the reservoir. The definition of reservoir boundaries can be heightened by study of seismic surveys, particularly 3-D surveys, and analysis of pressure buildups in wells after they have been brought on production.

The overall recovery of crude oil from a reservoir is a function of the production mech-anism, the reservoir and fluid parameters, and the implementation of supplementary recovery techniques. In general, recovery efficiency is not dependent upon the rate of production except for those reservoirs where gravity segregation is sufficient to permit segregation of the gas, oil, and water. Where gravity drainage is the producing mechanism, which occurs when the oil column in the reservoir is quite thick and the vertical permeability is high and a gas cap is initially present or is developed on pro-ducing, the reservoir will also show a significant effect of rate on the production ef-ficiency. Reservoir engineering expertise, together with geological and petrophysical engineering expertise, is being used to make very detailed studies of the production performance of crude oil reservoirs in an effort to delineate the distribution of resid-ual oil and gas in the reservoir, and to develop the necessary technology to enhance the recovery.

Well testing broadly refers to the diagnostic tests run on wells in petroleum reservoirs to determine well and reservoir properties. The most important well tests are called pressure transient tests and are conducted by changing the rate of a well in a prescribed way and recording the resulting change in pressure with time.

The information obtained from pressure transient tests includes estimates of:

1. Unaltered formation permeability to the fluid(s) produced in the well;

2. Altered (usually reduced) permeability near the well caused by drilling and completion practices;

3. Altered (increased) permeability near the well created by deliberately stimulating the well by injecting either an acid that dissolves some of the formation or a high-pressure fluid that creates fractures in the formation;

4. Distances to flow barriers located in the area drained by the well; and

5. Average pressure in the area drained by the well. In addition, some testing programs may confirm hypothesized models of the reservoir, including important variations of formation properties with distance or location of gas/oil, oil/water, or other fluid/fluid contacts.

Pressure transient tests are usually interpreted by comparing the observed pressure-time response to the predicted response by a mathematical model of the well/reservoir system. Graphical techniques are used to calculate permeability. More sophisticated graphical techniques involve matching changes in pressure to preplotted analytical solutions (type-curve matching). Regression analysis is used to match observed pressure-time data to mathematical models. Although analytical solutions are being found for more and more complex reservoir models each year, many reservoirs are still so complex that their behavior cannot be described accurately by analytical solutions. In such cases, finite-difference approximations to the governing flow equations can be used in commercial reservoir simulators, the reservoir properties treated as unknowns, and properties found that fit the observed data well.

Reservoir behavior can be simulated using models that have been constructed to have properties similar either to an ideal geometric shape of constant properties or to the shape and varying properties of a real (nonideal) oil or gas reservoir.

For application to petroleum reservoirs, it is necessary to predict the simultaneous flow behavior of more than one fluid phase having different properties (water, gas, and crude oil). The permeability, the relative permeability, and the density and viscosity of each phase constitute its transport properties for calculating its flow. The relative permeability is a factor for each phase (oil, water, gas) which, when multiplied by the permeability for a single phase such as water, will give the permeability for the given phase. It varies with the volume fraction of the pore space occupied by the phase, called the saturation of the given phase. Generally, the relative permeability of the water phase depends only on its own saturation, and likewise for the gas phase. The relative permeability of the oil phase is a function of the saturations of both gas and water phases.

Petroleum Trap

Petroleum trap is an underground rock formation that blocks the movement of petroleum and causes it to accumulate in a reservoir that can be exploited. The oil is accompanied always by water and often by natural gas; all are confined in a porous and permeable reservoir rock, which is usually composed of sedimentary rock such as sandstones, arkoses, and fissured limestones and dolomites. The natural gas, being lightest, occupies the top of the trap and is underlain by the oil and then the water. A layer of impermeable rock, called the cap rock, prevents the upward or lateral escape of the petroleum. That part of the trap actually occupied by the oil and gas is called the petroleum reservoir.

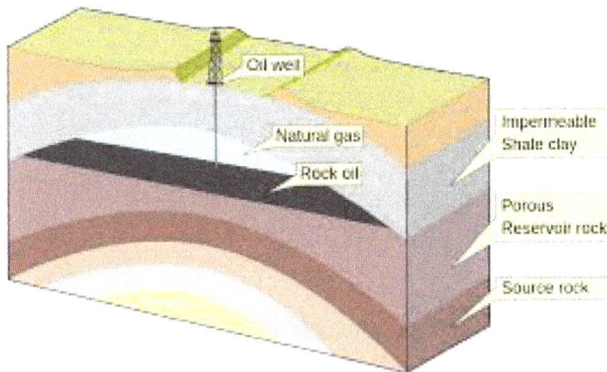

Many systems have been proposed for the classification of traps; one simple system divides them into structural traps and stratigraphic traps. The most common type of structural trap is formed by an anticline, a structure with a concave roof caused by the local deformation of the reservoir rock and the impermeable cap rock. In this case, the intersection of the oil-water contact with the cap rock determines the edges of the reservoir. Another kind of structural trap is the fault trap. Here, the fracture and slippage of rock along a fault line may bring an impermeable stratum in contact with a layer of permeable reservoir rock and thus forms a barrier to petroleum migration.

In a stratigraphic trap, variations within the rock strata themselves (e.g., a change in the local porosity and permeability of the reservoir rock, a change in the kinds of rocks laid down, or a termination of the reservoir rock) play the important role. The stratigraphic variations associated with the reservoir rocks are the main influence on the areal extent of the reservoirs in these traps.

The oil and gas pool will rise to the top of the trap if the underlying water is stationary, and the resulting oil-water contact will be level. When the water is moving, however, the pool is displaced down the trap's side in the direction of flow because of hydrodynamic pressure. In some traps, the pool may be displaced great distances or may even be completely flushed out.

Conventional Reservoirs

Conventional gas reservoirs contain 'free' gas in interconnected pore spaces that can flow easily to the wellbore i.e. natural flow is possible. In conventional natural gas reservoirs, the gas is often sourced from organic-rich shales which has migrated to these nearby sandstone or carbonate reservoirs, over geologic time.

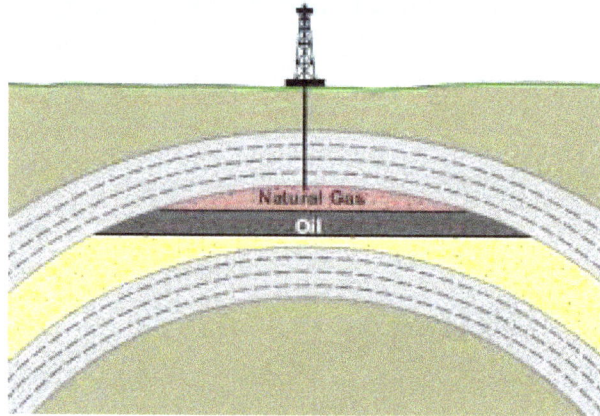

Formation of Conventional Reservoirs

Under very high pressures from overlying rocks, the light liquid oil is squeezed out of its source rock in which it was formed.

Subsequently, the oil penetrates into surrounding rocks through a system of interconnected fractures and pores. Oil always tends to move upwards due to its low density differential pressure decreasing towards towards the surface. Going up and upper through strata overlying its source rock, crude oil often takes a winding path as it seeks a place wherein an equilibrated pressure will stop to push it upward. It seeks the surface of the ground.

In the absence of any impediments, the oil will eventually seep out to the surface to form, upon volatilization of its lighter components, extremely dense deposits of wax and native asphalt. This, however, occurs on an exceptional basis.

Rocks overlying the source of oil are thousands of meters extremely diverse. Some of them are impervious. In the absence of interconnected pores or fractures, rocks act like a cork that prevents any continued migration of oil. Moreover, if the rock structure prevents lateral movement the oil is trapped.

The so-called anticlines are the structures that typically form oil traps. Anticlines are rock strata that have been bent by the lateral stress with their central parts moved upward to form a saddle-like structure.

Unconventional Reservoirs

Unconventional reservoirs are essentially any reservoir that requires special recovery operations outside the conventional operating practices. Unconventional reservoirs include reservoirs such as tight-gas sands, gas and oil shales, coalbed methane, heavy oil and tar sands, and gas-hydrate deposits. These reservoirs require assertive recovery solutions such as stimulation treatments or steam injection, innovative solutions that must overcome economic constraints in order to make recovery from these reservoirs monetarily viable. To help improve well economics in unconventional reservoirs it is important to evaluate wellbore architecture - including the cement sheath - as integral to well performance and total recovery.

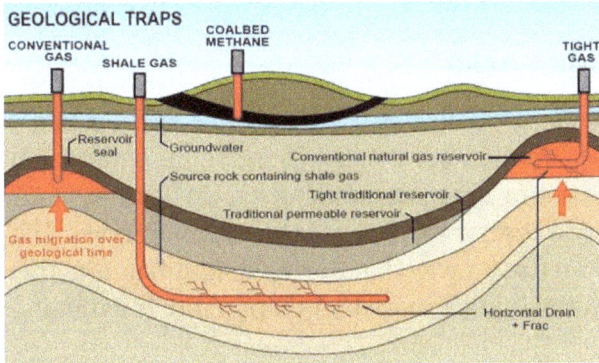

Reservoir Rock

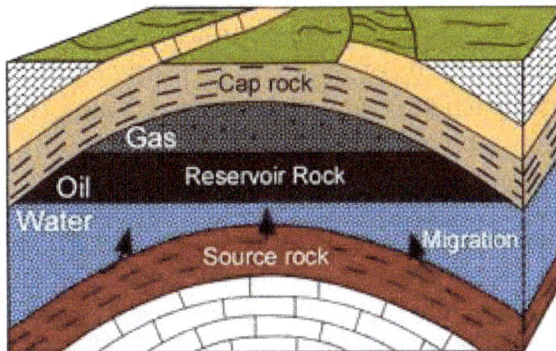

A reservoir rock is a rock containing porosity, permeability, sufficient hydrocarbon accumulation and a sealing mechanism to form a reservoir from which commercial flows of hydrocarbons can be produced. Porosity and permeability are the reservoir rock most significant physical properties. A fundamental property of a reservoir rock between them is porosity. However, for explorationists, an effective reservoir rock, the most fundamental reservoir rock property is its permeability. Both of them are geometric properties are the result of its lithological, structural and compositional behavior (composition). These physical compositions of a rock and the textural properties are geometric such as sizes and shapes of the rock grains, their arrangement system and packaging.

The reservoir content is estimated by means studying rock properties which can be determined in a direct way or indirectly. The indirect are done through laboratory measurements on core samples of reservoir rock of interest which constitutes direct methods of porosity data acquisition. This is done by measuring a bulk and their pore (empty spaces in a rock). Its bulk volume is gravimetrically determined when a core-sample is having an irregular shape. A petroleum system may have one or more reservoir rocks, and reservoir rocks may have different property basing on their types. Types of reservoir rock depend on kinds of their contents, composition, morphology and sedimentology.

Types of Reservoir Rocks

As a rock to be named a reservoir has to be a porous and permeable lithological structure. It encompasses sedimentary rocks. These sedimentary rocks may be made of sandstones (quartz sand or arksosic sandstone), carbonates mud or dolomite. Dolomites mostly form good reservoirs because the common reason behind it is that there is Mg, 13% smaller than Ca in a way that during dolomitization, there is a total decrease in volume of the material by 13%, here by 13% porosity is gained.

Sandstone Reservoir Rocks

The term sand refers to a specific grain with sizes between (62 μm - 2 mm). The performance of the sandstone as a reservoir rock is described by its combination of porosity and permeability depending on the degree to which the sand dominates its. The favorable texture is depicted by packaging of similar sized grains, not a combination of coarse and fine grained composition. The best sandstone reservoirs are those that are composed mainly of quartz grains of sand size of nearly equal sizes or silica cement, with minimal fragmented particles. Sandstone reservoirs are generally 25 meters thick.

Carbonate Reservoir Rocks

The most fascinating aspects of carbonate reservoir rocks are their content. Carbonates are usually made of fossils which "range from the very small single cell to the larger shelled animals". Most carbonate rocks are deposited at or in very close neighborhood to their site of creation. The "best-sorted" carbonate rocks are Oolites in which encompass grains of the same size and shapes even though Oolites are poorly sorted.

Properties of Reservoir Rock

   A. Porosity of reservoir is the property that tells how porous a rock is. It is also defined as a measure of the capacity of reservoir rocks to contain or store fluids. The porosity is genetically classified basing on standard sedimentologic description of reservoir rock; there are primary and secondary porosity.

The primary porosity types are:

i) Inter-particle: In this type by which rock content was quickly lost in muds and carbonate sands through compaction and cementation respectively. This type is mostly found as siliciclastic sands.

ii) Intra particle porosity by which the porosity is made of interiors of carbonate skeletal grains.

    a. Secondary porosity, the porosity formed after deposition leads to other couple of reservoirs types.

        i) Dissolution porosity type is made of carbonate dissolution and leaching. It is also called carbonate reservoirs.

        ii) Fracture porosity which is characterized by not being voluminous.

        Porosity can also be classified basing on rock morphology. There are three types of morphologies to the pore spaces which are:

         • Caternary in which the pore open to more than one throat passage

         • Cul-de-sac in which the pore open to only one throat passage

         • Closed pore in which there is no connection with other pores.

B. Permeability is a measure of the ability of a fluid to pass through its porous medium. Permeability is one of important to determine the effective reservoir. Porosity and permeability are two properties describing the reservoir rock capacity with regard to the fluid continence. Moreover, a reservoir rock can be porous without being permeable. For example it is said to be permeable if and only if the pores "communicate". Hence for explorationists, knowing reservoir rock permeability is a key mile stone because it is important for being used to determine if it really has sufficient commercial accumulation of oil, indeed measuring it is very difficult. The measuring of permeability can differently be understood basing on two different ways. When the porous medium is completely saturated by a single fluid, the permeability will be described ***absolute,*** become described as effective permeability when its porous medium is occupied by more than one fluid.

Other Factors Affecting the Volume of the Reservoir Rocks

1. Grain size and pattern arrangement: Apart from the arrangement pattern of grains size which effect rock properties, the actual size of the grains does not affects the permeability of a neither reservoir rock nor porosity.

2. Shape of the grains: grains with high sphericity tend to pack themselves well to make a minimum pore space, the fact which increases angularity and hence pore space volume increases.

3. Sorting or uniformity of size of the grains: size of grains has an effect on reservoir properties; the more uniform the grains are sized, the great proper volume of voids spaces. Thereby mixing grains of different sizes tends to decrease total volume of void space.

4. Subsequent action to the sediments (compaction): The more grains are compacted, more the volume of void spaces decreases. However the compaction of sand is less effective than the way clay does.

5. How the grains were formed.

Compressibility of Reservoir Rocks

The compressibility of reservoir rock is a factor which is generally neglected in reservoir engineering calculations. This is due in part to the fact that there is little published information on rock compressibility values for limestones and sandstones. Omission of rock compressibility is undoubtedly justified in calculations for saturated reservoirs; however, in undersaturated reservoirs, expansion of the rock accompanying decline in the reservoir pressure may be of such magnitude as to affect materially the prediction of reservoir performance. The effect of rock compressibility will be of most importance in:

(1) Calculation of oil in place by pressure decline data in undersaturated volumetric reservoirs when the limits of the field are unknow non indefinite,

(2) Studies of natural water drive performance. To estimate its importance in such cases, a series of laboratory tests were made to obtain usable values for reservoir rock compressibility.

Sandstone Reservoir Rocks

The term sand refers to a particular grain size (62 μm - 2 mm), not to a particular composition. The performance of the sandstone as a reservoir rock, its combination of porosity and permeability, depends upon the degree to which it is truly a sand. Texture should reflect similar sized grains, not a combination of coarse and fine grained material. The best sandstone reservoirs are those that are composed primarily of quartz grains of sand size, silica cement, with minimal fragmented particles.

The quality of the initial sandstone reservoir is a function of the source area for the materials, the depositional process, and the environment in which the deposition took place. Sandstone reservoirs are generally 25 meters thick, are lenticular and linear spatially, and less than 250 km² in area. They range in age from the oldest being Cambrian (in Algeria) to the youngest being Pliocene (Caspian region in Ukraine). In the USA, two-thirds of the sandstone reservoirs are Cenozoic in age.

Sandstone reservoirs form extensive stratigraphic traps. Many of the sandstones change composition within the sandstone unit; at this point, they are said to pinchout. Others represent ancient upland river channels that have left sand in the channel that has been converted into sandstone. Often these sand channels are stacked one on top of another which means that hydrocarbons are free to migrate between reservoirs. Other sandstone reservoirs are deltaic in origin, meaning they represent ancient river deltas similar to the Mississippi River delta that extends into the Gulf of Mexico. And successive sandstone layers exist in the subsurface that were created in alternating marine and terrestrial environments. So sandstone reservoir rocks are the result of a number of varied processes that can occur on dry land as well as beneath the sea.

Carbonate Reservoir Rocks

Carbonate rocks produce about 40% of all oil and gas, and include many of the reservoirs in Western Canada and the huge reservoirs in the Middle East. Carbonate reservoirs differ from sandstone reservoirs in several respects:

- Carbonate minerals are more soluble than silicate minerals; formation of secondary porosity is generally more important than in sandstones;

- Build-ups (bioherms, including reefs) produce upward projecting structures that may become (stratigraphic) hydrocarbon traps;

- Original mineralogy and diagenetic evolution are largely a function of biological conditions during deposition and prevailing marine hydrochemistry (i.e. aragonite vs. calcite seas);

- Carbonate rocks that otherwise would have low porosity and permeability commonly form fracture reservoirs;

- Carbonate minerals have different surface properties from silicate minerals and are often more oil wetting than sandstones.

To assess potential reservoirs during exploration, it is important to reconstruct the paleogeography as closely as possible, particularly the orientation of shoreline deposits (e.g., oolite sands) and distribution of reefs.

Porosity in Carbonate Reservoirs

Primary porosity in carbonate rocks consists mainly of:

1. Interparticle porosity in grainstones (e.g., between ooids, pellets, lithoclasts and fossils).

2. Intraparticle porosity in fossils (e.g., gastropods; bivalves).

3. Protected cavities under fossils or intraclasts (shelter porosity).

4. Cavities formed in carbonate mud due to gas bubbles or decayed microbial mats (fenestral porosity).

5. Primary cavities in reefs or coralline algae beds (growth framework porosity).

Secondary porosity can be formed through:

1. Chemical breakdown of minerals that are unstable in diagenetic pore fluids at any depth.

2. Biological breakdown - cavities formed by boring organisms, (e.g., bivalves, sponges).

3. Fracturing.

A major cause of secondary porosity is dissolution of aragonite through percolation of freshwater. If calcite is precipitated when aragonite is dissolved, a neomorphic transformation may occur, producing little secondary porosity. If pore fluids are undersaturated with respect to both aragonite and calcite, cavities will form. Aragonite shells may dissolve to give (secondary) moldic porosity. Timing is critical - an important factor is whether oil migrates into the pore before it fills with cement. Development of this type of porosity is controlled by the flow of meteoric water along the margins of and below sedimentary basins. Periods of uplift of land or lowering of sea level also cause extensive early dissolution (karstification) by meteoric water.

Another main type of secondary porosity is due to dolomitization. During dolomitization, the amount of dolomite precipitated may be less than that of the dissolved calcite or aragonite. The result may be a net increase in porosity. Some dolomitization is often associated with mixing zones of fresh and marine porewater, so these types of reservoirs commonly lie near the margin of the continental shelf, where meteoric porewater flows down into the basin (especially during periods of lowered sea-level).

Some porosity production takes place at depth due to maturation of organic matter. Expulsion of CO_2 and H_2S from shales, and decarboxylation, can lead to production of secondary porosity. Fracturing at depth, including that associated with over pressuring, may create new porosity and significantly enhance permeability for hydrocarbon migration.

Reduction of Porosity

Porosity in carbonate rocks is lost mainly by pressure solution and cementation. Well-sorted carbonate sand may undergo pressure solution at point contacts, and may precipitate cement between the grains. Precipitation of limited amounts of early carbonate cement, as in some beachrocks, causes the pressure to be more evenly distributed, and reduces pressure solution with burial. Consequently, carbonate sandstones that have been subjected to early diagenesis (minor cementation) may retain their remaining porosity best with depth.

With burial and increasing temperatures, the solubility of $CaCO_3$ declines, so late cementation (often large poikilotopic crystals) may take place, occluding both primary and secondary porosity. Consequently, predicting likely porosity conditions in carbonate rocks can be difficult.

Carbonate Reservoir Types

Although there are many specific types of carbonate reservoir, most fall with the following major groups:

Grainstones with Preserved (or Enhanced) Primary Porosity

This includes oolites, some pellet rocks, and those produced from accumulations of coarse shell debris. In geometry, they typically form prograding sheets or linear bars. Oolites can have primary porosities of ~40%, reaching 70% where mixed with shell debris.

Survival of primary porosity is favored in two general settings:

- Arid environments: Where there is little meteoric water near the land surface to produce cementation or neomorphic transformation of aragonite to calcite.

- Where early marine phreatic cements prop open and bind grains, but do not occlude the porosity.

Prospective sites are areas that lacked fluid flow, had early stagnant pore waters, or experienced an early influx of hydrocarbons. There are many examples —in arid settings, prograding oolite sheets are commonly overlain by shelf (subtidal and intertidal) muds, and then evaporites (anhydrite and gypsum) that can form a caprock.

Carbonate Slope Deposits

This group includes deposits on submarine slopes, commonly in the deeper waters seaward of carbonate platforms. The carbonates are mainly mass-flow deposits (carbonate turbidites, grainflow deposits, debris flows, etc.), including lithified reef rocks from platform margins and many other lithologies. These deposits may be coarse and may retain some primary porosity. Lenses and beds of pelagic carbonates may be interbedded with the coarser carbonate facies. The Tamabra facies of Poza Rica, Mexico is a well-known example.

Chalks

True chalks, such as the Upper Cretaceous chalks of northwest Europe and the more impure chalks in midwestern North America, are composed mainly of coccoliths that accumulated as pelagic oozes in warm, clear waters. They form sheet like deposits of great lateral extent. Although fine grained (< 7 μm), the coccoliths are composed of calcite, so they are relatively stable during diagenesis. Forming in deeper water, there are less prone to early exposure to meteoric waters during periods of low sea-level. Initial porosities can be high (30–40%), but their permeability is typically low (pores < 10μm) and they have a very high surface area.

Despite their low permeability, chalks have produced oil in the North Sea (Ekofisk) particularly where they are fractured. Elsewhere, however, they act as caprocks.

Reefs

Reefs form major petroleum reservoirs worldwide, but are commonly complex in terms of their porosity and reservoir facies. They are favored because the reef core commonly develops primary framework porosity, which commonly reach 60–80%. Furthermore, that framework inhibits compaction. If sea level falls, reefs are commonly exposed to early meteoric diagenesis, including karstification.

Retention of the high primary porosity depends large on the amount of fluid flow experienced early after the reefs have formed. If low, the porosity may endure. In contrast, some reefs lose primary porosity due to marine phreatic cementation, or exposure of meteoric waters saturated in $CaCO_3$. The history of the reef is critical. If the reef undergoes subsidence or is buried by transgressive marine shales (which may become a caprock), much of the primary porosity may survive.

The framework of the reef core is not always the best reservoir facies. In some examples, the reef rubble zone that surrounded the reef core may have higher porosity. Many examples of reef reservoirs are found in Western Canada, particularly in Devonian strata.

Stratigraphic Traps in Shelf Cycles

Stratigraphic traps commonly develop in prograding carbonate shelf cycles in sequences on many different ages, particularly those forming in arid settings (sabkha sequences). As the shoreline moves seaward, supratidal sediments come to overlie intertidal and shallow subtidal deposits. These produce sheets of low-energy carbonate muds that are commonly broken by higher energy deposits (channels, beaches, etc.) either normal or parallel to the shoreline.

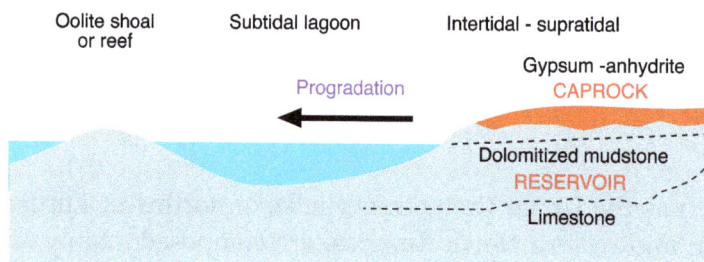

Primary porosity in the muds may be low, but it can be enhanced by early dolomitization (reflux or mixing with meteoric water) or karstification during periods of relatively lower sea-level. The sequence produced consists of evaporites (anhydrite) or mixed carbonate-evaporites that may become a caprock, dolomitized mudstones (commonly peloidal), and partially dolomitized peloidal limestones. Reservoir porosity may be present in either or both of the muddy carbonate units, and the associated coarser carbonate facies.

Paleokarstic Reservoirs — Carbonates below Unconformities

In humid environments, CO_2–charged rainwater and runoff may produce extensive weathering of exposed carbonate rocks, especially where organic soils have developed. This can generate much secondary porosity, including moldic, vuggy and even cavernous pores over very large areas.

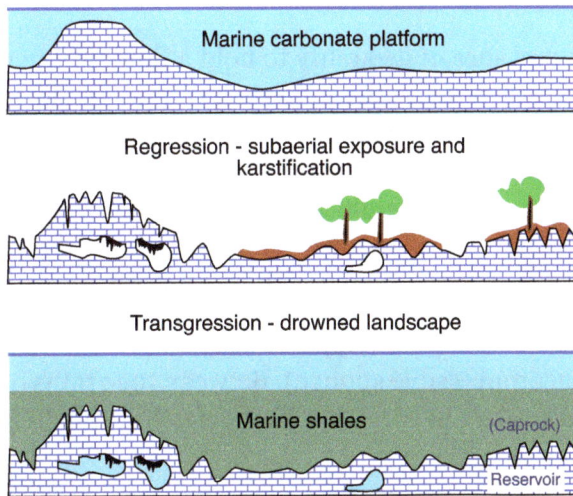

This is most likely to occur following a drop in sea-level or uplift of formerly submerged platforms. If the platform is later submerged during marine transgression, shales or evaporites may be deposited upon the unconformity surface to form a regional caprock.

Dolomite and Dolostone Reservoirs

Many reservoirs produce from dolostones of diverse origins. Dolomitization of carbonates (especially mudstones) commonly results in inter crystalline porosity with euhedral rhombs. If the porosity is not occluded by later precipitation or overgrowths, those pores may become a viable reservoir. Some dolostone reservoirs have porosities > 30%.

As with other carbonate reservoirs, the relative timing of porosity production and oil migration is critical. Dolostones have many origins, so generalizations are difficult. Some reservoirs are produced early, following local or regional reflux or mixing-zone dolomitization. Others are associated with burial dolomitization and late fracturing.

Porosity and Permeability

Porosity and permeability are related properties of any rock or loose sediment. Most oil and gas has been produced from sandstones. These rocks often have high porosity, and are usually "high perm" too. Porosity and permeability are absolutely necessary to make a productive oil or gas well. The petroleum geologist must stay focused on the porosity and permeability of the prospective reservoir.

- Porosity consists of the tiny spaces in the rock that hold the oil or gas.

- Permeability is a characteristic that allows the oil and gas to flow through the rock.

You need both porosity and permeability to make a producing oil or gas formation.

Porosity

Porosity of a rock is a measure of its ability to hold a fluid. Mathematically, porosity is the open space in a rock divided by the total rock volume (solid + space or holes). Porosity is normally expressed as a percentage of the total rock which is taken up by pore space. For example, a sandstone may have 8% porosity. This means 92 percent is solid rock and 8 percent is open space containing oil, gas, or water. Eight percent is about the minimum porosity that is required to make a decent oil well, though many poorer (and often non-economic) wells are completed with less porosity.

Even though sandstone is hard, and appears very solid, it is really very much like a sponge (a very hard, incompressible sponge). Between the grains of sand, enough space exists to trap fluids like oil or natural gas. The holes in sandstone are called porosity (from the word "porous").

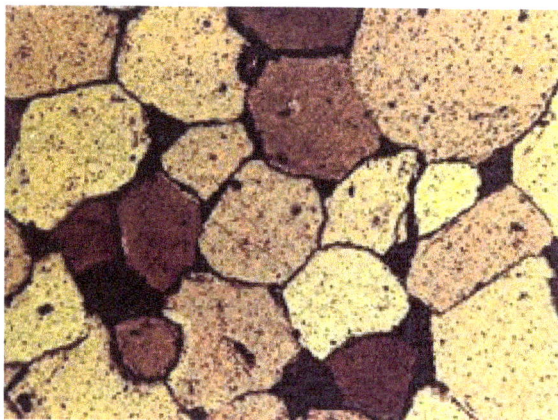

Sandstone under microscope

Here is a very thin slice (thinner than a human hair) of actual sandstone as seen through a microscope. The larger brown and yellow pieces are grains of "quartz," an extremely common mineral. Between the grains, you can see the porosity in the rock.

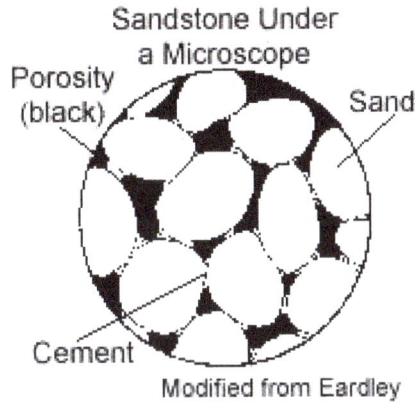

Sandstone Under
a Microscope

Porosity
(black)

Sand

Cement

Modified from Eardley

If you take a piece of sandstone and pour water on it, you will see the water is absorbed right into the rock. The water is soaked into the porosity.

The porosity is shown as black in the drawing on the right. Oil or gas will fill these holes in the rock.

Notice that the more spherical the grains are, the more space or porosity is left between them. Hence, a well-rounded sandstone will have more porosity than a poorly-routed one. A geologist loves to encounter a well-rounded sandstone, because they hold the most oil and gas of any of the clastic rocks.

Oil and gas are almost always found in such tiny spaces within rock pores. There is no "ocean" of oil underground.

Permeability

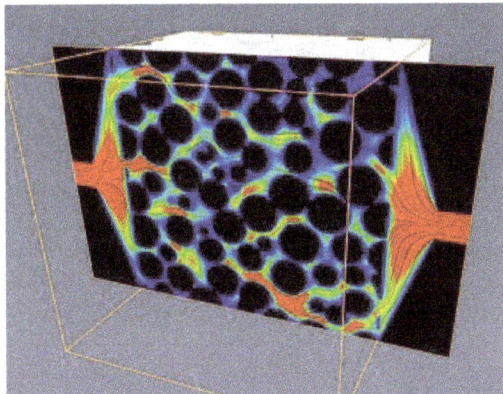

Permeability measures how easily fluid passes through a rock.

The permeability of a rock is a measure of the resistance to the flow of a fluid through a rock. If it takes a lot of pressure to squeeze fluid through a rock, that rock has "low permeability" or "low perm." If fluid passes through the rock easily, it has "high permeability," or "high perm."

Permeability Chart for Typical Sediments

Permeability	Pervious		Semi-Pervious		Impervious	
Unconsolidated Sand & Gravel	Well Sorted Gravel	Well Sorted Sand or Sand & Gravel	Very Fine Sand, Silt, Loess, Loam			
Unconsolidated Clay & Organic			Peat	Layered Clay	Unweathered Clay	
Consolidated Rocks	Highly Fractured Rocks		Oil Reservoir Rocks	Fresh Sandstone	Fresh Limestone, Dolomite	Fresh Granite
κ (cm²)	0.001 | 0.0001 | 10⁻⁵ | 10⁻⁶	10⁻⁷	10⁻⁸ | 10⁻⁹ | 10⁻¹⁰ | 10⁻¹¹	10⁻¹² | 10⁻¹³	10⁻¹⁴ | 10⁻¹⁵	
κ (millidarcy)	10⁺⁸ | 10⁺⁷ | 10⁺⁶ | 10⁺⁵	10,000	1,000 | 100 | 10 | 1	0.1 | 0.01	0.001 | 0.0001	

Permeability Chart

Permeability in petroleum-producing rocks is usually expressed in units called millidarcys (one millidarcy is 1/1000 of a darcy). Throughout drilling history, most oil and gas was produced from rocks that had ten to several hundred millidarcys. One darcy (1000 millidarcys) is a huge amount of permeability.

However, in the last 15 years, an increasing amount of US gas and oil production is coming from wells completed in shale formations. Shale actually has a lot of porosity (often much more than sandstone, may be 30% or so), but extremely low permeability due to the tiny grain size, which reduces the paths that the hyrdocarbons can follow. That means shale has historically been a poor producer of hydrocarbons. Gas has been produced through drilling from shales for well over a hundred years (gas needs less permeability to move through rock than oil), but quantities were small. Very few shale formations produced oil. Two things have changed the situation, allowing for increased shale oil and gas development. These two newer discoveries have allowed petroleum companies to artificially induce more permeability into petroleum-bearing rocks composed mainly of shale:

1. Horizontal Drilling: The widespread use of horizontal drilling technology, in which the drill bit is made to turn from the vertical to the horizontal (a 90-degree turn), where it can continue to drill horizontally through the formation. The horizontal track can be over a mile. *This exposes* much more *rock section than can be produced.*

2. Advances in Hydraulic Fracturing ("fracking"): "Fracking" is not a new technology — it has been around well over 70 years (despite what you might hear on the news). However, advances in fracking *techniques in* horizontally-drilled holes, particularly in shale formations, has led to a tremendous increase in shale oil and gas production. These new techniques allow the oil and gas operator to render low-permeability shale reservoirs more permeable, by artificially introduc-

ing small fractures into the formation. Hydrocarbons will readily flow through these artificially-induced fractures, vastly increasing the production from shale wells.

Reservoir Porosity

The perscentage of pore volume or void space or that volume within a rock that contains fluid is called porosity. A porous rock consists of particles and the spaces (pores) between them. Porosity is the ratio between the volume of the pores and the total volume.

Porosity = Volume of pores / Total volume of the rock

Voids ratio is also related to porosity but it is not the same, because voids ratio is the volume of the pores divided by the volume of particles.

Porosity is relatively easy to measure in the laboratory if a core can be recovered (core analysis). Porosity in-situ can be estimated by lowering down a well logging tool in the well.

The effective porosity of rocks varies between less than 1% and over 40%.

The porosity may often state this way

Porosity < 5% – LOW

Porosity 5% – 10 % – MEDIOCRE

Porosity 10%-20% MEDIUM

Porosity 20% – 30% GOOD

Porosity >30% – EXCELLENT

Porosity decreases with increase in depth. A high porosity is obviously desirable if the pores contain lots of oil or gas, but it may be economic to produce from a formation with a porosity as low as 4%. A porosity cut off issued to decide whether or not complete a well. In sandstone the typical cutoff is 8%.

Two types of porosities can exist in a rock. These are termed primary porosity and secondary porosity. Primary porosity is described as the porosity of the rock that formed at the time of its deposition. Secondary porosity develops after deposition of the rock. Secondary porosity includes vugular spaces in carbonate rocks created by the chemical process of leaching, or fracture spaces formed in fractured reservoirs. Porosity is further classified as total porosity and effective porosity. Total porosity is defined as the ratio of the entire pore space in a rock to its bulk volume. Effective porosity is the total porosity less the fraction of the pore space occupied by shale or clay. In very clean sands, total porosity is equal to effective porosity. As shown in figure, effective porosity represents pore space that contains hydrocarbon and non-clay water. Free formation water that is neither bound to clay nor to shale is called non-clay water. An accurate definition of effective porosity is total porosity minus volume of clay-bound water. The relationship between total porosity and effective porosity can be represented for a shaly sand model as:

$$\phi_t = \phi_e + V_{sh} \times \phi_{sh}$$

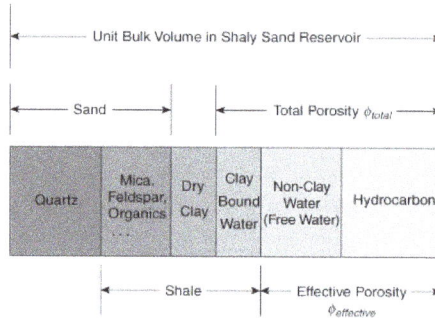

Figure: Porosity model for a shaly sand reservoir.

In equation $\phi_t = \phi_e + V_{sh} \times \phi_{sh}$, ϕ_t = total porosity, fraction; ϕ = effective porosity, fraction; V_{sh} = volume of shale, fraction; and ϕ_{sh} = shale porosity, fraction. The determination of shale porosity from well logs can be difficult and erroneous because the selection of the 100% shale section can be wrong and subjective. For this reason, an approximate form of equation $\phi_t = \phi_e + V_{sh} \times \phi_{sh}$ is obtained by replacing shale porosity ϕ_{sh} with total porosity ϕ_t to get:

$$\phi_t = \phi_e + V_{sh} \times \phi_t$$

For a clay model, effective porosity is represented as:

$$\phi_t = \phi_e + V_{cbw}$$

In equation $\phi_t = \phi_e + V_{cbw}$, V_{cbw} = volume of clay-bound water, fraction. The application of equation $\phi_t = \phi_e + V_{cbw}$ for calculation of accurate effective porosity depends on accurate quantification of the volume of clay-bound water. This can be determined from an elemental capture spectroscopy (ECS) well logs.

Reservoir Permeability

The permeability of a reservoir is a measure of how much fluid can flow through a rock for a specified pressure drop. Thus, permeability is a property of the rock, and is independent of the fluid (provided the rock is 100% saturated with that fluid). This means that the absolute permeability of a rock will be the same, whether the fluid is gas, oil, or water. What will change between these three fluids is the flow rate per unit pressure drop due to the different viscosities.

In a petroleum reservoir, the rock is usually not fully saturated with a single phase fluid. Generally saturations in the reservoir rock will consist of different amounts of gas, oil, and water. These saturations will change the effective permeability of the rock.

Permeability can be measured in a laboratory from core analysis. While this is some-times done with the core in its native or restored state, the more common method is to clean and dry the core, and measure its absolute permeability (usually to air or nitrogen, but the same value would be obtained if water were used instead). Permeability can also be determined by pressure transient analysis (PTA). However, it must be remembered that the permeability determined from analysis is the in-situ effective permeability of the primary reservoir fluid type and not the absolute permeability. The in-situ effective permeability is usually significantly less than the absolute core-derived permeability by a factor ranging from 2 to 200, depending on the reservoir. In addition, when a permeabil-ity is determined from a pressure transient test, it reflects the average permeability of the reservoir within the radius of investigation of the test (often several hundred feet). This is in contrast to a core measurement which represents only a few inches of the reservoir.

In PTA, effective permeability (in-situ) can be determined either by semi-log (radial) analysis or modeling (matching) pressure data. The results of these two techniques should be consistent, any inconsistencies should be accounted for by reviewing and modifying the information and data provided, and or modifying the analysis/model accordingly.

When gas, oil, and water are being produced during a test, the effective permeability for each phase can be estimated by assuming that only that fluid phase was flowing.

Note that in most of the pressure transient equations, the permeability term occurs as a mobility (k/μ) or a transmissivity (kh/μ) term.

Porosity vs. Permeability

- If solid rock completely surrounds a water-filled pore, then the water cannot flow.
- For groundwater to flow, pore spaces must be interconnected.
- The ability of a rock to allow a fluid to flow through an interconnected network of pores is called *Permeability*.
- If a rock has a high porosity, it does not necessarily have a high permeability. The pores must have interconnected conduits!
  - E.g. porous cork, is nearly impermeable
- Permeability depends on:
  - Number of available conduits
  - Size of conduits
  - Straightness of conduits

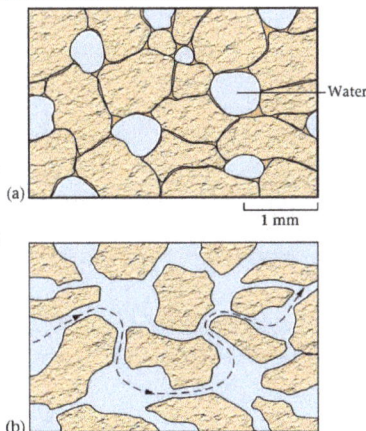

(a)

Water

1 mm

(b)

References

- Oil-formed-part-3-oil-reservoir: zionoil.com, Retrieved 16 March 2018

- Three-types-of-oil-reservoirs-9892: petropedia.com, Retrieved 16 March 2018

- Petroleum-reservoir-engineering: encyclopedia2.thefreedictionary.com, Retrieved 24 April 2018

- Petroleum-trap, science: britannica.com, Retrieved 11 June 2018

- Conventional-reservoirs: theprojectdefinition.com, Retrieved 28 April 2018

- 3-migration-or-origin-conventional-reservoirs: infolupki.pgi.gov.pl, Retrieved 14 May 2018

- Porosity-and-permeability-2: geomore.com, Retrieved 16 March 2018

Chapter 2

Properties of Crude Oil

An oil well predominantly produces crude oil and some natural gas dissolved in it. The hydrocarbons present in crude oil include alkanes, aromatic hydrocarbons and cycloalkanes, besides others. This unique mix of molecules determines the physical and chemical properties of petroleum, such as refractive index, viscosity, hydrogenation, paraffin wax content, etc. The topics elucidated in this chapter cover some of the important physical and chemical properties of crude oil.

Crude Oil

Crude oil, commonly known as petroleum, is a liquid found within the Earth comprised of hydrocarbons, organic compounds and small amounts of metal. While hydrocarbons are usually the primary component of crude oil, their composition can vary from 50%-97% depending on the type of crude oil and how it is extracted. Organic compounds like nitrogen, oxygen, and sulfur typically make-up between 6%-10% of crude oil while metals such as copper, nickel, vanadium and iron account for less than 1% of the total composition.

Crude Oil Formation

Crude oil is created through the heating and compression of organic materials over a long period of time. Most of the oil we extract today comes from the remains of prehistoric algae and zooplankton whose remains settled on the bottom of an Ocean or Lake. Over time this organic material combined with mud and was then heated to high temperatures from the pressure created by heavy layers of sediment. This process, known as diagenesis, changes the chemical composition first into a waxy compound called kerogen and then, with increased heat, into a liquid through a process called catagenesis.

Step 1: Diagenesis forms Kerogen

Diagenesis is a process of compaction under mild conditions of temperature and pressure. When organic aquatic sediments (proteins, lipids, carbohydrates) are deposited, they are very saturated with water and rich in minerals. Through chemical reaction, compaction, and microbial action during burial, water is forced out and proteins and carbohydrates break down to form new structures that comprise a waxy material known as "kerogen" and a black tar like substance called "bitumen". All of this occurs within the first several hundred meters of burial.

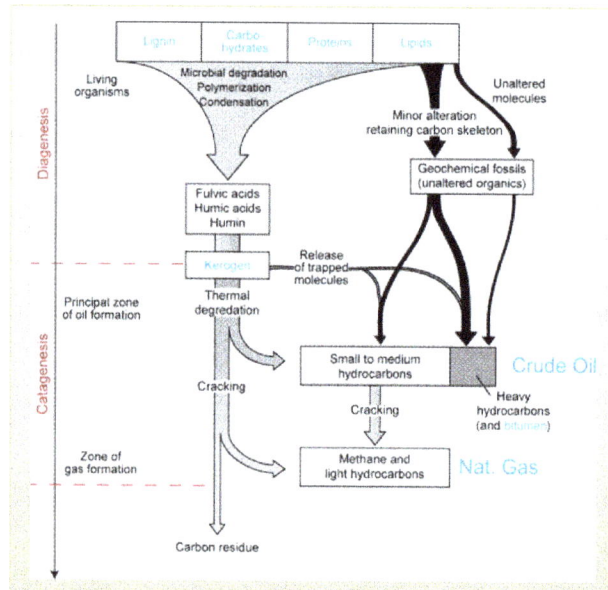

The bitumen comprises the heaviest components of petroleum, but the kerogen will undergo further change to make hydrocarbons and, yes, more bitumen.

Step 2: Catagenesis (or "Cracking") Turns Kerogen into Petroleum and Natural Gas

As temperatures and pressures increase (deeper burial) the process of catagenesis begins, which is the thermal degradation of kerogen to form hydrocarbon chains. Importantly, the process of catagenesis is catalyzed by the minerals that are deposited and persist through marine diagenesis. The conditions of catagenesis determine the product, such that higher temperature and pressure lead to more complete "cracking" of the kerogen and progressively lighter and smaller hydrocarbons. Petroleum formation, then, requires a specific window of conditions; too hot and the product will favor natural gas (small hydrocarbons), but too cold and the plankton will remain trapped as kerogen.

This behavior is contrary to what is associated with coal formation. In the case of terrestrial burial, the organic sediment is dominated by cellulose and lignin and the fraction of minerals is much smaller. Here, decomposition of the organic matter is restricted in a different way. The organic matter is condensed to form peat and, if enough temperature (geothermal energy) and pressure is supplied, it will condense and undergo catagenesis to form coal. Higher temperatures and pressures, in general, lead to higher ranks of coal.

Finding Petroleum

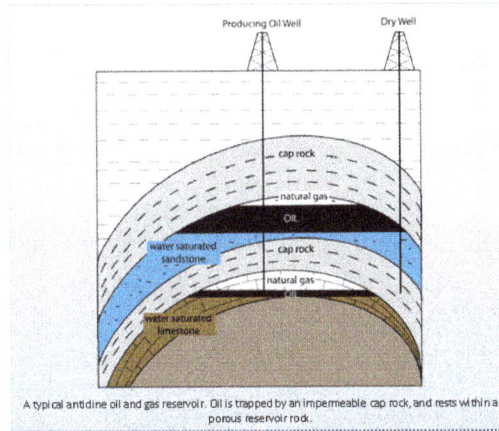

A typical antidine oil and gas reservoir. Oil is trapped by an impermeable cap rock, and rests within a porous reservoir rock.

Because the earth is filled entirely by layers of solid (or at significant depths) molten rock, the petroleum it contains cannot exist within a self-contained "lake", but must decide to live within the small fraction of space (or pores) that exist in these rocks. Like the sponge in your kitchen sink (albeit, less spongy and a bit heavier) certain kinds of rock (mainly sandstone and limestone) contain pores large enough and with enough connections to serve as storage and migration sites for oil or water or any other fluid wishing to call them home. Because most hydrocarbons are lighter than water and rock, those that exist within the earth will tend to migrate upwards until they reach the surface, or are trapped by an impermeable layer of rock.

There is a particular window of temperature that the zooplankton must find to form oil. If it is too cold, the oil will remain trapped in the form of kerogen, but too hot and the oil will be changed (through "thermal cracking") into natural gas. Therefore, the formation of an oil reservoir requires the unlikely gathering of three particular conditions: first, a source rock rich in organic material (formed during diagenesis) must be buried to the appropriate depth to find a desirable window; second, a porous and permeable (connected pores) reservoir rock is required for it to accumulate in; and last a cap rock (seal) or other mechanism must be present to prevent it from escaping to the surface. The geologic history of some places on earth makes them much more likely to contain the necessary combination of conditions.

Uses of Petroleum

To be of use to us, the crude oil must be "fractionated" into its various hydrocarbons. This is done at the refinery.

Oil can be used in many different products, and this is because of its composition of many different hydrocarbons of different sizes, which are individually useful in different ways due to their different properties. The purpose of a refinery is to separate and purify these different components. Most refinery products can be grouped into three classes: Light distillates (liquefied petroleum gas, naphtha, and gasoline), middle distillates (kerosene and diesel), and heavy distillates (fuel oil, lubricating oil, waxes, and tar). While all of these products are familiar to consumers, some of them may have gained fame under their refined forms. For instance, naphtha is the primary feedstock for producing a high octane gasoline component and also is commonly used as cleaning solvent, and kerosene is the main ingredient in many jet fuels.

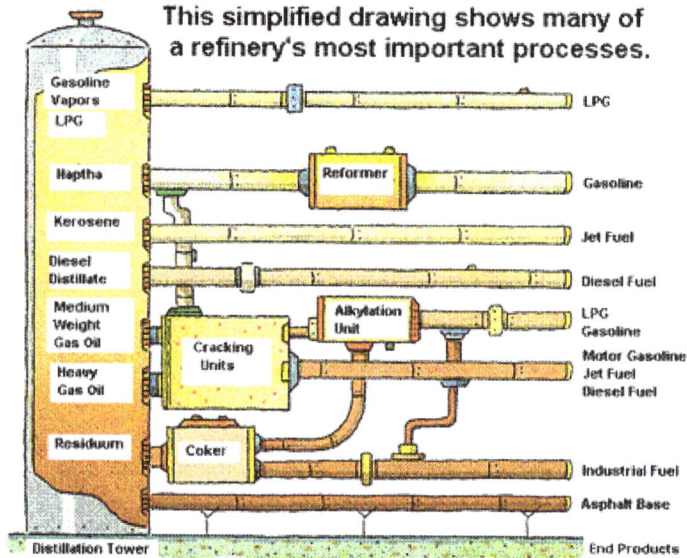

This simplified drawing shows many of a refinery's most important processes.

In a refinery, components are primarily separated using "fractional distillation". After being sent through a furnace, the crude petroleum enters a fractionating column, where the products condense at different temperatures within the column, so that the lighter components separate out at the top of the column (they have lower boiling points than heavier ones) and the heavier ones fall towards the bottom. Because this process occurs at atmospheric pressure, it may be called atmospheric distillation. Some of the heavier components that are difficult to separate may then undergo vacuum distillation (fractional distillation in a vacuum) for further separation. The heaviest components are then commonly "cracked" (undergoing catagenesis) to form lighter hydrocarbons, which may be more useful. In the same manner that natural mineral catalysts help to transform kerogen to crude oil through the process of catagenesis, metal catalysts can help transform large hydrocarbons into smaller ones. The modern form of "catalytic cracking" utilizes hydrogen as catalyst, and is thus termed "hydrocracking". This is a primary process used in modern petroleum refining to form more valuable lighter fuels from heavier ones. All of the products then undergo further refinement in different units that produce the desired products.

Alkanes are saturated hydrocarbons with between 5 and 40 carbon atoms per molecule which contain only hydrogen and carbon. The light distillates range in molecular composition from pentane (5 carbons: C_5H_{12}) to octane (8 carbons: C_8H_{18}). Middle distillates range from nonane (9 carbons: C_9H_{20}) to hexadecane (16 carbons: $C_{16}H_{34}$) while anything heavier is termed a heavy distillate. Hydrocarbons that are lighter than pentane are considered natural gas or natural gas liquids (liquefied petroleum gas).

A few further refinement processes are described below:

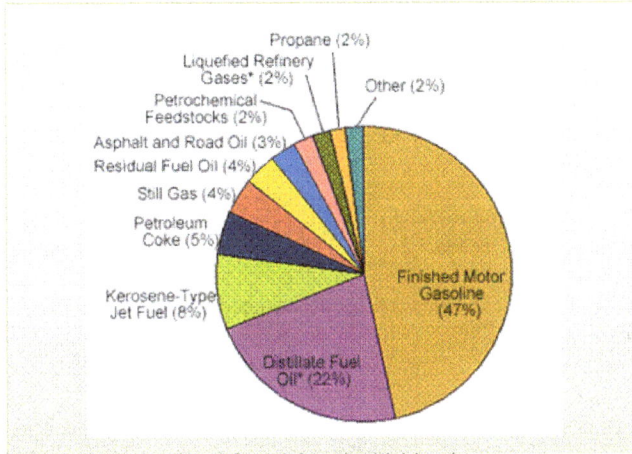

- Desalting removes salt from crude oil before entering fractional distillation.

- Desulfurization removes sulfur from compounds, and several methods are possible. Hydrodesulfurization is the typical method, and uses hydrogen to extract the sulfur. This occurs after distillation.

- Cracking breaks carbon-carbon bonds to turn heavier hydrocarbons into lighter ones. This can occur thermally (as occurs during the petroleum formation process beneath the earth) or through the action of a catalyst:

  ○ Thermal Cracking

    ▪ Steam, visbreaking, or coking

  ○ Catalytic cracking

    ▪ Fluid catalytic cracking (FCC) cracks heavy oils into diesel and gasoline. Uses a hot fluid catalyst.

    ▪ Hydrocracking (similar to FCC but lower temperature and using hydrogen as catalyst) cracks heavy oils into gasoline and kerosene.

- A catalytic reformer converts naphtha into a higher octane form, which has a higher content of aromatics, olefins, and cyclic hydrocarbons. Hydrogen is a byproduct, and may be recycled and used in the naphtha hydrotreater.

- Steam reforming is a method of producing hydrogen from hydrocarbons, which may then be used in other processes.

- Solvent dewaxing removes heavy wax constituents from the vacuum distillation products.

Coal to Liquids Technology

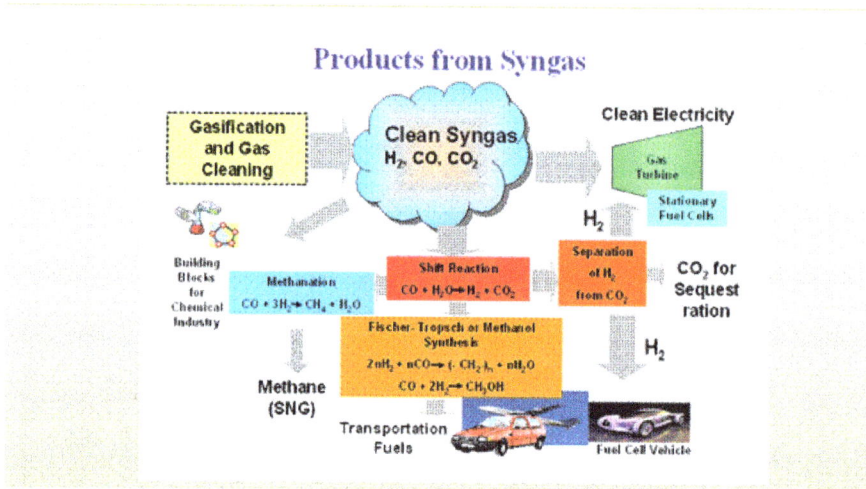

Products from Syngas

But what if we want the same fuels that we get from petroleum, without the petroleum? Is there another way? Actually, yes, we can use coal. The only commercial coal to liquids (CTL) industry in operation today is in South Africa, where coal-derived fuels have been in use since 1955, and currently account for about 30% of the country's gasoline and diesel consumption.

There are two different methods for converting coal into liquid fuels, direct and indirect liquefaction.

- The direct liquefaction method dissolves the coal in a solvent at high temperature and pressure. While highly efficient, the liquid products generated this way require further refining to achieve a high fuel grade.

- The indirect liquefaction technique gasifies the coal to form a "syngas" (a mixture of, primarily, hydrogen and carbon monoxide produced by breaking down the coal into its components using high temperature and pressure with the injection of steam and oxygen). This gas is then condensed over a catalyst (in the "Fischer-Tropsch" process) to produce a higher quality, cleaner fuel. Syn-fuel processes (such as Fisher-Tropsch) actually build up larger hydrocarbons from smaller ones, which is the opposite of cracking.

Both of these methods result in the release of carbon dioxide in a proportion greater than that produced during the extraction and refinement of petroleum, but the fuels they produce may be cleaner than final petroleum fuels. Carbon dioxide sequestration has been proposed as a method to counteract this downside, thereby achieving cleaner fuels, without the drawback of carbon dioxide release.

Physical Properties of Crude Oil

The physical properties of crude oils are the quantitatively measurable characteristics of crude oils. They vary according to the composition of the oil, the relative abundance of the groups of hydrocarbons, and essentially depend on reservoir temperatures and pressures.

Specific (or A.P.I) Gravity

API stands for the American Petroleum Institute, which is the major United States trade association for the oil and natural gas industry. The API represents about 400 corporations in the petroleum industry and helps to set standards for production, refinement, and distribution of petroleum products. They also advocate on behalf of the industry. One of the most important standards that the API has set is the method used for measuring the density of petroleum. This standard is called the API gravity.

Specific gravity is a ratio of the density of one substance to the density of a reference substance, usually water. The API gravity is nothing more than the standard specific gravity used by the oil industry, which compares the density of oil to that of water through a calculation designed to ensure consistency in measurement. Less dense oil or "light oil" is preferable to more dense oil as it contains greater quantities of hydrocarbons that can be converted to gasoline.

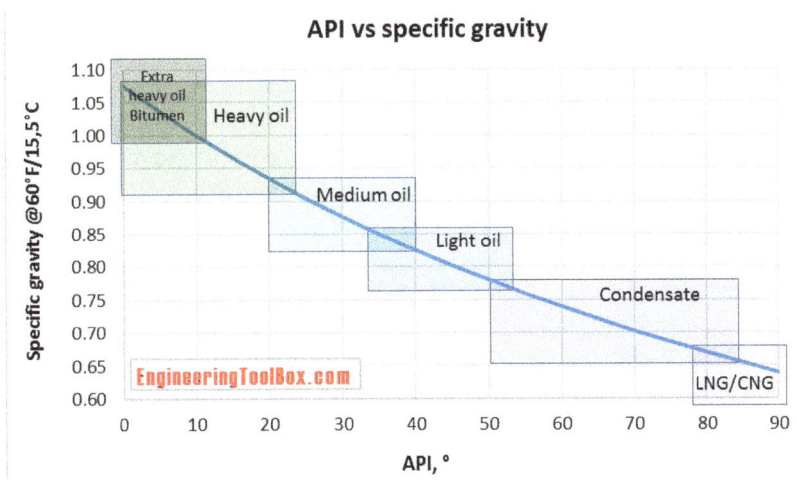

Petroleum is less dense that water and in 1916, the U.S. government instituted the Baumé scale as the standard measure for any liquid less dense than water. This, in most cases, applies to oil. The value used in this scale was 141.5, but subsequent testing showed that, due to error, the actual value should be 140. The government changed the scale to 140 to correct the issue, but the use of 141.5 had become so entrenched in the oil industry that the API decided to create the API gravity scale using the old value of 141.5.

API gravity is calculated using the specific gravity of oil, which is nothing more than the ratio of its density to that of water (density of the oil/density of water). Specific gravity for API calculations is always determined at 60 degrees Fahrenheit. API gravity is found as follows:

$$\text{API gravity} = (141.5/\text{Specific Gravity}) - 131.5$$

Though API values do not have units, they are often referred to as degrees. So the API gravity of West Texas Intermediate is said to be 39.6 degrees. API gravity moves inversely to density, which means the denser oil is, the lower its API gravity will be. An API of 10 is equivalent to water, which means any oil with an API above 10 will float on water while any with an API below 10 will sink.

The API gravity is used to classify oils as light, medium, heavy, or extra heavy. As the "weight" of oil is the largest determinant of its market value, API gravity is exceptionally important. The API values for each "weight" are as follows:

- Light – API > 31.1

- Medium – API between 22.3 and 31.1

- Heavy – API < 22.3

- Extra Heavy – API < 10.0

These are only rough valuations as the exact demarcation in API gravity between light and heavy oil changes depending on the region from which oil came. The fluctuation as to what constitutes light crude in a given region is the result of commodity trading in oil.

Because density is a measure of weight per volume, API can be used to calculate how many barrels of crude can be extracted from a metric ton of a given oil. A metric ton of West Texas Intermediate, with an API of 39.6, will produce 7.6 barrels (at 42 gallons each). The calculation is:

$$\text{Barrels per metric ton} = 1/[(141.5/(\text{API} + 131.5) \times 0.159]$$

Viscosity

The viscosity of a crude oil is the property that describes its resistance to movement, mainly due to collisions, electrostatic forces and hydrogen bonding between its molecules that are moving in different velocities and directions. It is an important property used in reservoir modeling, design of production and transportation equipment, as well as in the design of processing facilities in crude oil refineries. Additionally, in the case of an accidental oil spill, viscosity plays an important role on the impact and the fate of the oil spill and consequently on the response measures. Modeling the weathering processes of an oil spill requires knowledge of the viscosity of the fluid. As weathering progresses, the composition as well as the physical properties of the fluid, including viscosity are changing constantly and a method is required to estimate viscosity since initial values are no longer valid. The incentive of this study was to develop an accurate method of estimating crude oil viscosity that would provide the necessary input to such oil spill weathering model. Crude oils are complex mixtures consisting of thousands of components of different type such as aliphatic, naphthenic and aromatic hydrocarbons, as well as other polar compounds containing heteroatoms (nitrogen, sulphur and oxygen). Therefore a purely theoretical prediction of crude oil viscosity is extremely difficult.

There are numerous published predictive models and correlations for the estimation of the liquid (kinematic) viscosity of crude oils and hydrocarbon mixtures. Typically they can be categorized into two groups, theoretical models or empirical correlations. Theoretical models are usually more accurate but they are more complex and require a better characterization of the oil including parameters such as critical properties, chemical composition, acentric factor or other similar properties which in many cases are not available. Empirical correlations are typically simpler and only require a few basic parameters such as mid-boiling point temperature and specific gravity of the oil, or alternatively a single viscosity measurement, in order to extrapolate the viscosity to a different temperature. Scientists have shown that the effect of temperature on viscosity can be correlated accurately by Walther's equation:

$$\ln = [\ln(v+0.8)] = a_1 + a_2 \ln(T)$$

where v = predicted kinematic viscosity (cSt), and T = viscosity prediction temperature (K).

When a single viscosity measurement is available at a specific tem-perature, the viscosity at a different temperature can be estimated byreplacing a_1 with the double logarithm of the known viscosity Scientists. In this case equation ($\ln = [\ln(v+0.8)] = a_1 + a_2 \ln(T)$)

becomes:

$$\ln = [\ln(v+0.8)] = \ln [\ln (v_0 +0.8)] + a_1 \ln(T/T_0)$$

where v = predicted kinematic viscosity (cSt), T = temperature of pre-diction (K), v_0 = measured kinematic viscosity (cSt) and T_0 = temperature of the viscosity measurement (K).

Using a complete set of known viscosities, Scientists estimated the value of a_2 to be equal to -3.7, predicted with an AAD equal to 1.47%. This factor has also been confirmed in the current study, with a small adjustment to be equal to -3.682.

The correlations presented in table that do not require a viscosity measurement instead require the 50% mass boiling point (T_b) as an input. In this study we have tested both the 50% mass boiling point as well as the weighted average T_b of the crude oil, since the latter describes the skewed distribution of boiling points in a crude oil better than the mid-point. The 50% mass boiling point is the temperature where 50% of the mass of the oil has been distilled. The weighted average T_b is determined by the relationship:

$$\text{weighted average} T_b = \sum_{i=I}^{n} \Delta x_i \times T_{bi}$$

where n = the number of distillation cuts covering the complete range of oil, Δx_i = the mass fraction of the distillation cut i with $\Sigma \Delta x_i$ = 1, and Tb_i = mid boiling point temperature of distillation cut i (K).

The calculation of the weighted average requires the completed istillation range to be known, and preferably the full True Boiling Point(TBP) distillation curve.

Refractive Index

Refractive index of crude oils is measured as a function of temperature from 20° C to 60° C in steps of 5° C. Also, refractive index of 1-Propanoal and 1-Butanol as a function of temperature is measured from 20° C to 32° C in steps of 2° C. The measurements have done using an Abbe Refractometer. The data are tabulated in table.

Table: mesured refractive index data at different temperatures

(t) ± 0.01 (°C)	Refractive index, $\pm 1 \times 10^{-5}$		(t) ± 0.01 (°C)	Refractive index, $\pm_1 \times 10^{-5}$	
	Heavy oil	Light oil		Propanol	Butanol
20	1.6668	1.4530	20	1.3861	1.3990
25	1.6639	1.4525	22	1.3856	1.3986
30	1.6597	1.4492	24	1.3852	1.3981
35	1.6565	1.4478	26	1.3849	1.3977
40	1.6535	1.4454	28	1.3844	1.3972
45	1.6498	1.4436	30	1.3840	1.3969
50	1.6459	1.4429	32	1.3837	1.3965
55	1.6434	1.4398			
60	1.6402	1.4379			

The refractive index is found to decrease linearly with increasing temperature for all the liquids. We have fitted n as a function of t as,

$$n_{heavy} = 1.68029 - 6.736 \times 10^{-4} t$$

$$n_{tight} = 1.46118 - 3.83967 \times 10^{-4} t$$

$$n_{Bu\,tan\,ol} = 1.40319 - 2.09107 \times 10^{-4} t$$

$$n_{Pro\,panol} = 1.39009 - 1.99643 \times 10^{-4} t$$

As t increase from 20 °C to 60 °C, n_{heavy} falls by about 3% and n_{light} decreases by about 1.5%. On the other hand, n of Propanol and Butanol decrease only by 0.24% when t is increased from 20 °C to 32 °C.

Optical Activity

Optical activity is an effect of an optical isomer's interaction with plane-polarized light.

Optical isomers, or enantiomers, have the same sequence of atoms and bonds but are different in their 3D shape. Two enantiomers are nonsuperimposible mirror images of one another (i.e., chiral), with the most common cited example being our hands. Our left hand is a mirror image of our right, yet there is no way our left thumb can be over our right thumb if our palms are facing the same way and placed over one another. Optical isomers also have no axis of symmetry, which means that there is no line that bisects the compound such that the left half is a mirror image of the right half.

Optical isomers have basically the same properties (melting points, boiling points, etc.) but there are a few exceptions (uses in biological mechanisms and optical activity). There are drugs, called enantiopure drugs that have different effects based on whether the drug is a racemic mixture or purely one enantiomer. For example, d-ethambutol treats tuberculosis, while l-ethambutol causes blindness. Optical activity is the interaction of these enantiomers with plane-polarized light.

Rotation of Light

An enantiomer that rotates plane-polarized light in the positive direction, or clockwise, is called dextrorotary [(+), or d-], while the enantiomer that rotates the light in the negative direction, or counterclockwise, is called levorotary [(-), or l-]. When both d- and l- isomers are present in equal amounts, the mixture is called a racemic mixture.

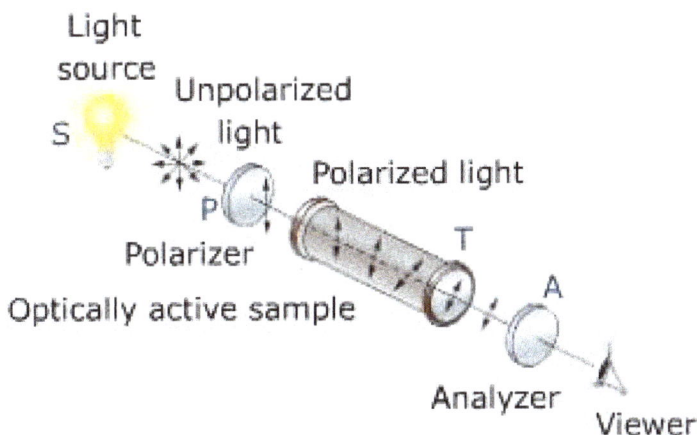

In the figure above, you can see that unpolarized light passes through a filter so that only waves that oscillate in a certain direction can pass through. When these waves interact with an optically active material, they are rotated either clockwise or counter-clockwise, depending on the enantiomer. In the case of the figure above, the light is rotated clockwise so the substance is the dextrorotary enantiomer.

Measuring Optical Activity

Optical activity is measured by a polarimeter, and is dependent on several factors: concentration of the sample, temperature, length of the sample tube or cell, and wavelength of the light passing through the sample. Rotation is given in +/- degrees, depending on whether the sample has d- (positive) or l- (negative) enantiomers. The standard measurement for rotation for a specific chemical compound is called the specific rotation, defined as an angle measured at a path length of 1 decimeter and a concentration of 1g/ml. The specific rotation of a pure substance is an intrinsic property. In solution, the formula for specific rotation is:

$$[\alpha]_{\lambda}^{T} = \frac{\alpha}{l \cdot c}$$

where

- $[\alpha]$ is the specific rotation in degrees cm³ dm⁻¹ g⁻¹,

- λ is the wavelength in nanometers,

- α is the measured angle of rotation of a substance,

- T is the temperature in degrees,

- l is the path length in decimeters,

- c is the concentration in g/ml.

Although oils and oil distillates are usually dextrorotatory, some levorotations can be measured in lower boiling distillates. Fractional distillation of a Colombia crude oil of Eocene age showed that optical rotation is maximal at an average molecular weight of 464. The rotatory power of petroleum reflects the extent to which its composition has been transformed during geochemical maturation. Optical rotations thus can be correlated with the degree of maturity in much the same way as can such other compositional features as aromatic character or content of steroid-like substances, which are demonstrable in petroleum.

The optical activity of petroleum is gradually degraded by maturational influences, and a decreasing trend in rotatory power is recognized in a comparison of oils produced from rocks of progressively greater geologic age. The optical rotation of dark crude oils can be measured in a polarimeter of suitable photometric sensitivity. The measurements extend geochemical application of oil polarimetry, which was restricted formerly to oil distillates and decolorized oils.

Cloud and Pour Points

The pour point of a liquid is the temperature at which it becomes semi solid and loses its flow characteristics. In crude oil a high pour point is generally associated with a high paraffin content, typically found in crude deriving from a larger proportion of plant material.

Two pour points can be derived which can give an approximate temperature window depending on its thermal history. Within this temperature range, the sample may appear liquid or solid. This peculiarity happens because wax crystals form more readily when it has been heated within the past 24 hrs and contributes to the lower pour point.

The upper pour point is measured by pouring the test sample directly into a test jar. The sample is then cooled and then inspected for pour point as per the usual pour point method.

The lower pour point is measured by first pouring the sample into a stainless steel pressure vessel. The vessel is then screwed tight and heated to above 100 °C in an oil bath. After a specified time, the vessel is removed and cooled for a short while. The sample is then poured into a test jar and immediately closed with a cork carrying the thermometer. The sample is then cooled and then inspected for pour point as per the usual pour point method.

Automatic Method

ASTM D5949, Standard Test Method for Pour Point of Petroleum Products (Automatic Pressure Pulsing Method) is an alternative to the manual test procedure. It uses automatic apparatus and yields pour point results in a format similar to the manual method (ASTM D97) when reporting at a 3 °C.

The D5949 test method determines the pour point in a shorter period of time than manual method D97. Less operator time is required to run the test using this automatic method. Additionally, no external chiller bath or refrigeration unit is needed. D5949 is capable of determining pour point within a temperature range of -57 °C to +51 °C. Results can be reported at 1 °C or 3 °C testing intervals. This test method has better repeatability and reproducibility than manual method D97.

Under ASTM D5949, the test sample is heated and then cooled by a Peltier device at a rate of 1.5 +/- 0.1°C/min. At either 1°C or 3°C intervals, a pressurized pulse of compressed gas is imparted onto the surface of the sample. Multiple optical detectors continuously monitor the sample for movement. The lowest temperature at which movement is detected on the sample surface is determined to be the pour point.

Importance of Pour Point

- Pour point is the temperature above which a lubricant or fluid will move freely under normal conditions.

- Oil and gas companies pay close attention to pour point because it has an impact on drilling and transport. If a petroleum deposit has a high pour point, usually reflecting a high paraffin content, it may be difficult to extract. Drilling teams need the oil to flow so they can pull it up with drilling rigs. Transport can also become a problem; in some cases, oil pipelines need to be heated to keep the oil at pour point and ensure it moves smoothly from oil fields to shipping terminals and other destinations.

- Manufacturers of lubricants also have concerns in this area. For products like motor oil, the lubricant may need to operate at a range of temperatures. Technicians don't want oil that flows too readily at low temperatures because it might thin too much at high temperatures and cause problems with the engine. They also need to consider issues like handling engines in extreme cold, where it is sometimes necessary to heat lubricants or entire engine blocks to keep the equipment operational.

Pour Point of Lubricating Oil and Base Oil

Type of Oil	Deg C (°)
Base Oil SN 150	-8
New Engine Oil	-8
Diesel	-1

Cloud Point

The cloud point of a fluid is the temperature at which dissolved solids are no longer completely soluble, precipitating as a second phase giving the fluid a cloudy appearance. This term is relevant to several applications with different consequences.

In the petroleum industry, cloud point refers to the temperature below which wax in diesel or bio wax in biodiesels form a cloudy appearance. The presence of solidified waxes thickens the oil and clogs fuel filters and injectors in engines. The wax also accumulates on cold surfaces (e.g. pipeline or heat exchanger fouling) and forms an emulsion with water. Therefore, cloud point indicates the tendency of the oil to plug filters or small orifices at cold operating temperatures.

In crude or heavy oils, cloud point is synonymous with wax appearance temperature (WAT) and wax precipitation temperature (WPT).

The cloud point of a nonionic surfactant or glycol solution is the temperature where the mixture starts to phase separate and two phases appear, thus becoming cloudy. This behavior is characteristic of non-ionic surfactants containing polyoxyethylene chains, which exhibit reverse solubility versus temperature behavior in water and therefore "cloud out" at some point as the temperature is raised. Glycols demonstrating this behavior are known as "cloud-point glycols" and are used as shale inhibitors. The cloud point is affected by salinity, being generally lower in more saline fluids the cloud point of a solution, whereas builders or other salts will depress the cloud point temperature.

Automatic Method

ASTM D5773, Standard Test Method of Cloud Point of Petroleum Products (Constant Cooling Rate Method) is an alternative to the manual test procedure. It uses automatic apparatus and has been found to be equivalent to test method D2500.

The D5773 test method determines the cloud point in a shorter period of time than manual method D2500. Less operator time is required to run the test using this automatic method. Additionally, no external chiller bath or refrigeration unit is needed. D5773 is capable of determining cloud point within a temperature range of -60 °C to +49 °C. Results are reported with a temperature resolution of 0.1 °C.

Under ASTM D5773, the test sample is cooled by a Peltier device at a constant rate of 1.5 +/- 0.1 °C/min. During this period, the sample is continuously illuminated by a light source. An array of optical detectors continuously monitor the sample for the first appearance of a cloud of wax crystals. The temperature at which the first appearance of wax crystals is detected in the sample is determined to be the cloud point.

The cloud point of petroleum products and biodiesel fuels is an index of the lowest temperature of their utility for certain applications. Wax crystals of sufficient quantity can plug filters used in some fuel systems.

Cloud Point of Lubricating Oil

Type of Oil	Deg C (°)
New Engine Oil	-15
Petroleum	-43
Diesel	6

Volume

The volume of a crude oil in its reservoir rock differs from the volume it occupies in the surface. This is due to formation gas-oil ratio and reservoir pressures. The formation gas-oil ratio expresses the volume of gas contained in one barrel of a crude oil as it comes from the reservoir rock. Under high reservoir pressure, the volume of oil in the reservoir increases because of the influence of dissolved gases. But on release of the reservoir pressures, the dissolved gases escape, leading to the shrinkage of the volume of the crude oil at the surface.

Flourescence

In the petroleum industry, the characterization of crude oil is a key factor for improved optimization of the refining process. Spectroscopic techniques have gained relevance for this purpose because of their potential to give rapid responses containing valuable information related to the intrinsic chemical characteristics of each analyzed sample.

Petroleum oils are mixtures of aliphatic, aromatic and high molecular weight organic compounds. Due to this heterogeneity, chemical analysis is generally complex. Petroleum oils are typically characterized using liquid chromatography, separating into saturates, aromatics, resins, and asphaltenes. Each of these components can be further characterized by the use of Gas Chromatography–Mass Spectrometry (GC–MS) which provides unambiguous identification of individual components.

Infrared (IR, both mid-IR, MIR, and near- IR, NIR) spectroscopies are the most widely used in spectroscopic techniques for crude oil analysis. The simplicity of sample handling and the rapidity of the analysis have facilitated the development of new ways to determine the physical chemical properties of petroleum products (gasoline, kerosene, diesel, etc.). In spite of its good signal/noise ratio, rapid response, simplicity and low cost, IR still presents several limitations, such us the overlapping of absorption bands and saturation of the signal due to the high absorbance of dense sample such as crude petroleum.

Another technique that is gaining importance in the petrochemical field is fluorescence spectroscopy, due to its sensitivity and selectivity. Fluorescence has been extensively used in the petroleum industry for the analysis and classification of different petroleum samples and reviews describing the basis of the technique and its inherent advantages and disadvantages can be found in the literature. Factors such as the specific chemical composition (concentration of fluorophores and quenching species) and physical (viscosity and optical density) influence emission intensity and wavelength. In the case of heavy oils, fluorescence emission is generally broad, very weak, and has short lifetimes, whereas lighter oils have narrower more intense emission bands, and longer lifetimes. The nature of the emission is governed by the complex interplay between energy transfer and quenching caused by the high concentrations of fluorophores and quenchers in petroleum oils. The complexity of crude oils usually prevents the resolution of any specific chemical component in terms of individual emission parameters.

Several different solvents can be used to dilute crude oil to obtain a solution transparent to transmit light. Nerveless organic solvents are highly flammable, depending on their volatility. Exceptions are some chlorinated solvents like dichloromethane and chloroform. Mixtures of solvent vapors and air can explode. Many solvents can lead to a sudden loss of consciousness if inhaled in large amounts is the case of chloroform. Nujol is a brand of mineral oil used in infrared spectroscopy. It is a heavy paraffin oil so it is chemically inert and has a relatively uncomplicated IR spectrum. It density is

0.838 g/mL at 25 °C. The empirical formula of Nujol is essentially the alkane formula $C_nH_{(2n+2)}$ where n is very large. Crude oil and nujol can be mixed in a uniform phase under vigorous shacking. Absorption and emission in the UV and visible range are well knowing and less intense than crude oil emissions. Moreover using excitation in the visible range it is possible to access only crude oil absorption and consequently only crude oil emission.

Remote sensing techniques, such as Light Detection and Ranging with lasers (Lidar) can be able to detect oil spills and even potential oil exploration areas. By the measurement of oil thickness the value of absorption coefficient for many samples and in a wide wavelength band (300– 600 nm) should be known ad hoc or be inferred at laboratory scale. Also by knowing the crude oil emission spectra can be possible to identify the oil origin and its quality.

Color

Crude oils contain hundreds of different molecules, and their compositions vary widely from one geographic location to another. The quality of the crude is largely a function of its maturation (not unlike the quality of coal). The color which can range from light yellow through green to almost black is usually indicative of the types of properties the crude will have.

The lighter the color of the crude, the greater is its maturation; the smaller the average number of carbon atoms in the hydrocarbon molecules are; and the more profitable it will be since less processing is needed to convert it into a desirable fuel source. Younger crudes tend to have greater percentages of large molecular weight molecules.

Odor

Different approaches can be applied to refinery odor problems such as instrumental measurements (e.g. gas chromatography (GC) and gas chromatography – mass spectrometry (GC-MS)), sensory methods (e.g. olfactometry (dilution technique)), odor panels, and hybrid instrumentation. Generally, odor measurement is necessary for odor regulation and control. Odor detectability or threshold or concentration, odor intensity, odor persistence, hedonic tone, odor character or quality, and annoyance are the terms associated with odor measurement. A sensory property referring to the minimum concentration that produces an olfactory response or sensation is called odor detectability or threshold or concentration. Odor concentration is measured as dilution ratios and given as Dilution to Threshold (D/T) (is a measure of the number of dilutions needed to make the odorous air non-detectable) and sometimes assign the pseudo-dimension of odor units per cubic meter (m^3). Odor unit (OU) is the concentration divided by the threshold. Odor intensity is the strength of the perceived odor sensation. The relationship between intensity and concentration can be explained by Stevens' law or the power law:

$$t=k(C)^n$$

Where, I, k, C, and n are intensity (parts per million of butanol), concentration, constant, and exponent ranges from about 0.2 to 0.8, depending on the odorant, respectively. Odor intensity scales of 1, 2, 3, 4, and 5 depict barely perceptible, slight, moderate, strong, very strong, respectively. Odor persistence describes the rate at which an odor's perceived intensity decreases as the odor is diluted. Hedonic tone is a measure of the pleasantness or unpleasantness of an odor. Odor character or quality refers to the property to identify an odor and differentiates it from another odor of equal intensity. Annoyance is defined as interference with comfortable enjoyment of life and property. Odor measurement can be done by several ways such as instrumental methods/chemical analysis, electronic methods and sensory test methods/olfactometry. Odor sensory methods, instead of instrumental methods, are normally used to measure it. Odor sensory methods are available to monitor odor both from source emissions and in the ambient air. Odor threshold of some compounds which may be found in refinery emissions have been given in table.

Table: Refinery odours and their possible sources

Type of smell	Odor compounds	Sources
Bad eggs	H_2S + trace of disulfides	Crude storage, distillation of gases, sulfur removal, flare stacks (cold flare)
Sewer smell	Dimethyl sulfide, ethyl and methyl mercaptans	Effluent water, biological treatment plants, LPG odorizing spent caustic loading and transfer
Burnt oil	Unsaturated hydrocarbons	Catalytic cracking unit, coking asphalt blowing, asphalt storage
Gasoline	Hydrocarbons	Product storage, American Petroleum Institute (API) and corrugated plate interceptor (CPI) separators
Aromatics (benzene) Hot tar	Benzene , toluene H_2S, mercaptans, hydrocarbons	Aromatic plants, Naphtha reformers Asphalt storage

Odour Emission Factors (OEF)

Odor emission factors (OEF) have been developed in analogy with the emission factors defined by the U.S. EPA that can be used to estimate the odor emission rate (OER) associated with an industrial plant, odor impact assessment, and etc. OER can be estimated as follows:

$$OER = A \times OEF \times \left(1 - \frac{ORE}{100}\right)$$

Where, OER is the odor emission rate (in ou_E/s); A denotes the activity index; and ORE is the overall odor reduction efficiency (%) that can be calculated using the following equation:

$$ORE = 100 \times \left(\frac{C_{od,IN} - C_{od,OUT}}{C_{od,IN}}\right)$$

Where, $C_{od,IN}$ and $C_{od,OUT}$ are the odor concentrations at the inlet and at the outlet of the abatement system, respectively.

Control of Odours in the Petroleum Refinery

The techniques which can be used to reduce or control the odour generation in the petroleum refinery are:

- The use of nitrate-based products in septic water areas (e.g. storage tanks, sewage systems, oil/water separators) in order to replace bacteria feedstock and to favor the development of denitrificative bacteria, which will both reduce added nitrates in nitrogen and existing hydrogen sulphide in sulphates;

- Reduction of volatile organic compounds (VOCs) and odor generation by covering some units of wastewater treatment plant (WWTP) (e.g. CPI and API separators) with closed sealed covers;

- Reduction of odors from water buffer tanks by maintaining the smallest possible surface area of oil and water in contact with air via using a fixed roof tank or a floating roof tank;

- Reduction and control of fugitive emissions;

- Control of flares and prevention or reduction of emissions from them;

- Control of fuel quality;

- The use of scrubbing systems for odorous gases; and

- The use of incineration systems for odorous gases.

Coefficient of Expansion

The Coefficient of Thermal Expansion is the rate change in the size of an object (could be solid, liquid or in gaseous state) with the rate change in the temperature. It generally measures the fractional change in the size of the object with the change in the temperature keeping the pressure constant. There are several types of thermal coefficients that are developed such as linear, volumetric and area thermal coefficients.

Solids undergo maximum expansion when their surface temperature is increased by heating while they also contract when they are cooled. This response to the temperature change by any object is referred as coefficient of thermal expansion.

Coefficient of thermal expansion is used in:

- Linear thermal expansion

- Volumetric thermal expansion

- Area thermal expansion

The thermal coefficient of expansion for crude oil and petroleum products is expressed by the following equation:

$$\alpha_{60} = \left(\frac{K_0}{\rho} + K_1 \right) \frac{1}{\rho*} + K_2$$

Where

K_0, K_1 and K_2 depends on the type of oil used.

Aqueous Solubility

Increase in temperature increased crude oil solubility, and the higher molecular weight species were affected more positively than lower molecular weight species. Increases in pressure or salinity decreased solubility. The presence of gas in solution increased the solubility of high molecular weight hydrocarbons ($> C_{24}$) over all temperatures, and increased the solubility of lower molecular weight hydrocarbons at high temperatures ($> 180–260 \,°C$). Gas decreased the solubility of low molecular weight hydrocarbons.

Surface Tension Effect

The effect of temperature on the interfacial tension between crude oil and gemini surfactant solution

Abstract: The effect of temperature on the interfacial tension between gemini surfactant solution and crude oil is investigated in this paper. The interfacial tension is measured by the spinning drop interfacial tension-meter of Texas-500c. For surfactants 14-4-14 and 16-4-16, the interfacial tensions are very sensitive to temperature and undergo minima with the increase of the temperature. Temperature has also an effect on the dynamic interfacial tension, i.e. increasing the temperature can shorten the time needed for an interfacial tension to reach equilibrium. However, for a mixing surfactant system, the effect of temperature on the interfacial tension between oil and surfactant solution is not remarkable because of synergism.

Flash Point

Flash point is the lowest temperature at which a liquid (usually a petroleum product) will form a vapor in the air near its surface that will "flash," or briefly ignite, on exposure to an open flame. The flash point is a general indication of the flammability or combustibility of a liquid. Below the flash point, insufficient vapor is available to support combustion. At some temperature above the flash point, the liquid will produce enough vapor to support combustion. (This temperature is known as the fire point.)

The use of the flash point as a measure of a liquid's hazardousness dates from the 19th century. Before gasoline became important, kerosene was the main petroleum product (used mainly as fuel for lamps and stoves), and there was a tendency on the part of petroleum distillers to leave as much as possible of the commercially worthless gasoline in the kerosene in order to sell more product. This adulteration of kerosene with highly volatile gasoline caused numerous fires and explosions in storage tanks and oil lamps. Legal measures were instituted to curb the danger, and test methods were prescribed and minimum flash points set.

Flash points are measured by heating a liquid to specific temperatures under controlled conditions and then applying a flame. The test is done in either an "open cup" or a "closed cup" apparatus, or in both, in order to mimic the conditions of storage and the workplace. Representative liquids and their approximate flash points are:

- Automotive gasoline, −43° C (−45° F)
- Ethyl alcohol, 13° C (55° F)
- Automotive diesel fuel, 38° C (100° F)
- Kerosene, 42−72° C (108−162° F)
- Home heating oil, 52−96° C (126−205° F)
- SAE 10W-30 motor oil, 216° C (421° F)

Commercial products must adhere to specific flash points that have been set by regulating authorities.

Chemical Properties of Crude Oil

The chemical properties of crude oil deal with the chemical nature and the changes in composition in relation to temperature and pressure variations occurring at all times within the oil pool. Some of the chemical properties are related to the origin, migration, and accumulation of the crude oil.

Chemical Nature

Petroleum (Crude oil) consists of mainly carbon (83-87%) and hydrogen (12-14%) having complex hydrocarbon mixture like paraffins, naphthenes, aromatic hydrocarbons, gaseous hydrocarbons (from CH_4 to C_4H_{10}). Table gives more details about composition of petroleum. Besides crude oil also contains small amount of non-hydrocarbons (sulphur compounds, nitrogen compounds, oxygen compounds) and minerals heavier crudes contains higher sulphur. Depending on predominance of hydrocarbons, petroleum is classified as paraffin base, intermediate base or naphthenic base.

Table: Composition of Petroleum

Hydrocarbons

Hydrogen Family	Distinguishing characteristics	Major hydrocarbons	Remarks
Paraffins (Alkanes)	Straight carbon chain	Methane, ethane, propane, butane, pentane, hexane	General formula C_nH_{2n+2} Boiling point increases as the number of carbon atom increases. With number of carbon 25-40, paraffin becomes waxy.
Isoparaffins (Iso alkanes)	Branched carbon chain	Isobutane, Isopentane, Neopentane, Isooctane	The number of possible isomers increases as in geometric progression as the number of carbon atoms increases.
Olefins (Alkenes)	One pair of carbon atoms	Ethylene, Propylene	General formula C_nH_{2n} Olefins are not present in crude oil, but are formed during process. Undesirable in the finished product because of their high reactivity. Low molecular weight olefins have good antiknock properties.
Naphthenes	5 or 6 carbon atoms in ring	Cyclopentane, Methyl cyclopentane, Dimethyl cyclopentane, cyclohexane, 1,2 dimethyl cyclohexane.	General formula C_nH_{2n+2}-2Rn RN is number of naphthenic ring The average crude oil contains about 50% by weight naphthenes. Naphthenes are modestly good components of gasoline.
Aromatics	6 carbon atom in ring with three around linkage.	Benzene, Toluene, Xylene, Ethyl Benzene, Cumene, Naphthaline	Aromatics are not desirable in kerosene and lubricating oil. Benzene is carcinogenic and hence undesirable part of gasoline.

Non Hydrocarbons

Non-hydrocarbons	Compounds	Remarks
Sulphur compounds	Hydrogen sulphide, Mercaptans	Undesirable due to foul odour 0.5% to 7%
Nitrogen compounds	Quinotine, Pyradine, pyrrole, indole, carbazole	The presence of nitrogen compounds in gasoline and kerosene degrades the colour of product on exposure to sunlight. They may cause gum formation normally less than 0.2.
Oxygen compounds	Naphthenic acids, phenols	Content traces to 2%. These acids cause corrosion problem at various stages of processing and pollution problem.

Hydrogenation of Crude Oil

Hydrogenation is a chemical reaction between molecular hydrogen and an element or compound, ordinarily in the presence of a catalyst. The reaction may be one in which hydrogen simply adds to a double or triple bond connecting two atoms in the structure of the molecule or one in which the addition of hydrogen results in dissociation (breaking up) of the molecule (called hydrogenolysis, or destructive hydrogenation). Typical hydrogenation reactions include the reaction of hydrogen and nitrogen to form ammonia and the reaction of hydrogen and carbon monoxide to form methanol or hydrocarbons, depending on the choice of catalyst.

Nearly all organic compounds containing multiple bonds connecting two atoms can react with hydrogen in the presence of a catalyst. The hydrogenation of organic compounds (through addition and hydrogenolysis) is a reaction of great industrial importance. The addition of hydrogen is used in the production of edible fats from liquid oils. In the petroleum industry, numerous processes involved in the manufacture of gasoline and petrochemical products are based on the destructive hydrogenation of hydrocarbons. In the late 20th century the production of liquid fuels by hydrogenation of coal has become an attractive alternative to the extraction of petroleum. The industrial importance of the hydrogenation process dates from 1897, when the French chemist Paul Sabatier discovered that the introduction of a trace of nickel as a catalyst facilitated the addition of hydrogen to molecules of carbon compounds.

The catalysts most commonly used for hydrogenation reactions are the metals nickel, platinum, and palladium and their oxides. For high-pressure hydrogenations, copper chromite and nickel supported on kieselguhr (loose or porous diatomite) are extensively used.

Paraffin Wax Content

Wax deposition is one of the chronic problems in the petroleum industry. The various crude oils present in the world contain wax contents of up to 32.5%. Paraffin waxes consist of straight chain saturated hydrocarbons with carbons atoms ranging from C18 to C36. Paraffin wax consists mostly with normal paraffin content (80–90%), while, the rest consists of branched paraffins (iso-paraffins) and cycloparaffins. The sources of higher molecular weight waxes in oils have not yet been proven and are under exploration. Waxes may precipitate as the temperature decreases and a solid phase may arise due to their low solubility. For instance, paraffinic waxes can precipitate out when temperature decreases during oil production, transportation through pipelines, and oil storage. The process of solvent dewaxing is used to remove wax from either distillate or residual feed stocks at any stage in the refining process. The solvents used, methyl-ethyl ketone and toluene, can then be separated from dewaxed oil filtrate stream by membrane process and recycled back to be used again in solvent dewaxing process.

Odd Carbon Chain Lengths

The different chain lengths have progressively higher boiling points, so they can be separated out by distillation. This is what happens in an oil refinery crude oil is heated and the different chains are pulled out by their vaporization temperatures.

The chains in the C_5, C_6 and C_7 range are all very light, easily vaporized, clear liquids called naphthas. They are used as solvents dry cleaning fluids can be made from these liquids, as well as paint solvents and other quick-drying products.

The chains from C_7H_{16} through $C_{11}H_{24}$ are blended together and used for gasoline. All of them vaporize at temperatures below the boiling point of water. That's why if you spill gasoline on the ground it evaporates very quickly.

Next is kerosene, in the C_{12} to C_{15} range, followed by diesel fuel and heavier fuel oils (like heating oil for houses).

Next come the lubricating oils. These oils no longer vaporize in any way at normal temperatures. For example, engine oil can run all day at 250 degrees F (121 degrees C) without vaporizing at all. Oils go from very light (like 3-in-1 oil) through various thicknesses of motor oil through very thick gear oils and then semi-solid greases. Vasoline falls in there as well.

Chains above the C_{20} range form solids, starting with paraffin wax, then tar and finally asphaltic bitumen, which is used to make asphalt roads.

All of these different substances come from crude oil. The only difference is the length of the carbon chains.

Carbon Isotope Ratio

Carbon isotopes have been used for many years in crude oil/crude oil and crude oil/source rock correlation work. The method is based on the fact that the carbon isotope ratios of crude oils, rock extracts (i.e., carbon compounds soluble in organic solvents), and kerogen (insoluble organic compounds) are similar if they are genetically related, i.e., if the kerogen is the organic source of the rock extract or of the crude oil.

Several more sophisticated isotopic correlation techniques have been developed recently, such as the "isotope type-curve" method. This method is an empirical approach which gives information on oil/oil or oil/source rock correlation and allows the identification of bacterial degradation of crude oils and rock extracts. Crude oils or rock extracts are separated into saturates, aromatics, heterocomponents and asphaltenes. The carbon isotope patterns of these fractions are used for oil/oil correlations and for the approximate estimation of the $^{13}c/^{12}c$ ratios of the source rock kerogen. The possibilities of oil/oil correlations can be considerably improved in many cases by the additional determination of the hydrogen isotope ratios of the oil fractions.

Gas/source rock correlations are applied even more directly in hydrocarbon exploration. The applications are based on the relationship between the carbon isotope ratios of the methane, and the type and maturity (i.e. vitrinite reflectance R^o) of the organic material from which the methane had been formed.

Typical applications are carbon isotope determinations of methane from:

- New gas reservoirs or gas shows, in order to identify their source rocks. This information can define targets for drilling operations and can influence the drilling strategy in frontier areas.

- Cuttings which will show if hydrocarbons present in the cuttings have been generated in-situ or not. These isotope determinations permit the identification of migrated hydrocarbons.

The method is presently being improved and modified for the isotopic identification of gases which leak from reservoirs to the surface and are absorbed in surface sediments.

Two main problems have been recognized:

1. The amounts of methane present in sediments are very small. A technique had to be developed for handling extremely small amounts of gas without introducing isotope fractionation by chemical or mass spectrometric procedures.

2. Changes in the hydrocarbon composition and isotope ratios can happen before, during, and after the sampling of the sediments. Procedures have been developed which allow identification of secondary isotope fractionation caused by oxidizing bacteria in the bottom sediments or by degassing during the storage of the samples.

Laboratory experiments have been carried through to overcome these difficulties, and the identification of deeply pooled hydrocarbons by isotope analyses of sediment gases will probably soon become a competitive tool in hydrocarbon exploration.

Sulphur and its Isotope Ratio

Crude oils and bitumens vary in sulfur content from less than 0.05 to more than 14%, although relatively few produced crude oils contain more than 4% sulfur. Most oils contain from 0.1 to 3% sulfur.

Hydrogen sulfide and elemental sulfur dissolved in crude oils usually are a very minor portion of the total sulfur if they are present at all

Studies of sulfur isotope ratios ($^{34}S/^{32}S$)* have contributed to our understanding of the origin of sulfur in petroleum and support the view for its early 'incorporation into oil precursor materials. These studies indicate that a large portion of the sulfur has been derived from reduced sulfur. H_2S and So) showing typical isotopic fractionation attributed to microbial reduction of sulfate %der conditions where the supply of sulfate is not largely reduced.

Porphyrins in Crude Oil

Porphyrin compounds and related pigments are good geochemical marker compounds in the genesis of petroleum. Porphyrins or chlorins, isoprenoids and perylenes exist in rocks of all ages from early in the Precambrian to the present, and in modern plants. Of the distinctive pigment compounds, porphyrins predominate in old rocks, chlorins in recent. Mass spectrometry reveals the existence of phyllo and etio homologous series of porphyrins and chlorins with mass ranges from about 408 to 504. Low sulfur low- vanadyl-porphyrin oils show a direct relationship between vanadyl porphyrins and sulfur in heavy petroleums. Separate origins for the porphyrins of vanadyl and nickel pigments are indicated, with vanadyl coming from vanadyl chlorins of plants, and nickel from chlorophyll degradation products. The surface activity role of porphyrin and chlorin pigments in oil field formation is uncertain. Evidence for molecular associates of porphyrin pigments may reveal distinctive geochemical behaviour of lipids or proteins during oil field genes.

Trace Metals in Crude Oils

The crude oils are mostly based on two elements carbon and hydrogen and almost all crude oil ranges from 82-87% carbon and 12-15% hydrogen. Crude oil contains three basic chemical series: paraffins, naphthenes, and aromatics. The crude oils from different sources may not be completely identical.

The paraffins are also called methane series, and comprises most common hydrocarbons in crude oil. The paraffins that are liquid at normal temperature boil between 40-200 °C. The naphthenes are saturated closed ring series and are important part of all liquid refinery products. The aromatics are unsaturated closed ring series. Benzene is most common of the series and is present in most of the crude oils, but aromatics constitute a small fraction of all crudes.

The crude oil also contains sulphur, nitrogen and oxygen in small quantities. Sulphur is the third most abundant constituent of crude oil. The total sulphur in crude oil varies from below 0.05% up to 5% or more. Generally greater the specific gravity of the crude oil, higher is its sulphur content. The oxygen contents of the crude oil are usually less than 2%. Nitrogen is present in most of the crude oils, usually in quantities of less than 0.1%.

Preliminary fractionation of crude oil according to chemical class is carried out before identification of individual components. Several such fractionation and isolation schemes are available depending on the type of crude oil under investigation. One of the separation scheme is based on SARA method, which has name from the fractions produced, namely saturates (S), aromatics (A) resins (R) and asphaltenes (A). The sample is adsorbed on the silica, or alumina, followed by the selective elution of the components with increasingly polar solvent, have reviewed the crude oil separation and identification including SARA method. HPLC and infra-red spectroscopy have also been used for SARA characterization.

Asphaltenes consist of polar fraction of the crude oil comprising polyaromatics, heteroaromatics and various metals.

Metal ions in Crude Oil

The metals present in the crude oils are mostly Ni(II) and VO(II) porphyrins and non-porphyrins. Other metal ions reported form crude oils, include copper, lead, iron, magnesium, sodium, molybdenum, zinc, cadmium, titanium, manganese, chromium, cobalt, antimony, uranium, aluminum, tin, barium, gallium, silver and arsenic. Metalloporphyrins are among the first compounds identified to belong to biological origin. proposed that plant chlorophylls transformed into the geoporphyrins. Metalloporphyrins in crude oils are of fundamental interest from geochemical context for better understanding geochemical origin of petroleum source. The information could be useful for catagenetic oil formation, maturation of organic matter, correlation, depositional and environmental studies. Vanadium and nickel metalloporphyrins are present in large quantity in heavy crude oils. Their presence cause many problems because such metals have a deleterious effect on the hydrogenation catalysts used in upgrading processes.

Among the porphyrins encountered in the crude oils, etioporphyrins (etio) and dexo-phylloerithroetioporphyrin (DPEP), and their homologues are more frequently observed.

Effect of Carbon Dioxide and Saline Water

High amount of anaerobic micro bacteria in some reservoir rocks generate carbon dioxide, which in association with hydrocarbon gases, constitute dissolved gases that mobilize the liquid hydrocarbon. The carbon dioxide facilitates a decrease in viscosity of the oil and generates internal gas pressure to drive the crude oil from dead-end pockets and through interstitial spaces.

When these gases are dissolved in saline water, sufficient reservoir drive to flow light paraffin base oils is provided. The saline nature of the water reduces its surface tension, thus, creating molecular contact between water and the crudes, subsequently, leading to the effective mobilization of the crude oil.

References

- What-Is-Crude-Oil-A-Detailed-Explanation-On-This-Essential-Fossil-Fuel: oilprice.com, Retrieved 30 May 2018

- Cloud-and-pour-point, laboratory-services: pentasflora.com, Retrieved 13 April 2018

- Coefficient-of-thermal-expansion-5360: petropedia.com, Retrieved 09 July 2018

- A-simple-and-flexible-correlation-for-predicting-the-viscosity-of-crude-oils-319420391: researchgate.net, Retrieved 16 March 2018

- Hydrogenation, science: britannica.com, Retrieved 25 June 2018

Chapter 3

Trap Formation

Millions of years of heat and pressure have led to the metamorphosis of microscopic plants and animals into hydrocarbons such as oil and natural gas. A trap is formed when buoyant forces that drive the upward migration of hydrocarbons fails to overcome the capillary forces of a sealing medium. Traps are classified into structural traps, stratigraphic traps and hydrodynamic traps. This chapter closely examines some of the crucial aspects of these different trap formations.

Structural Traps

Structural trap is a type of geological trap that forms as a result of changes in the structure of the subsurface, due to tectonic, diapiric, gravitational and compactional processes. These changes block the upward migration of hydrocarbons and can lead to the formation of a petroleum reservoir.

Structural traps are the most important type of trap as they represent the majority of the world's discovered petroleum resources. The three basic forms of structural traps are the anticline trap, the fault trap and the salt dome trap.

Anticlinal Trap

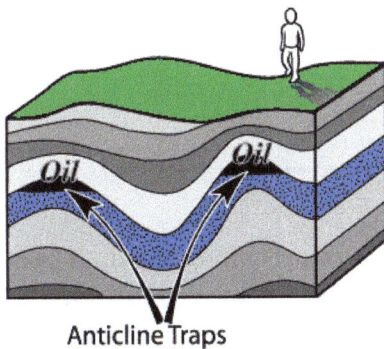

Anticline Traps

An anticline trap is created by the upfolding of rocks, similar to an arch. Oil moves to the highest point in this arch's dome and then comes to rest In order for this trap to be effective, there needs to be a rock above the dome in order to seal the oil in place This type of structural trap was first discovered by early geologists in the late nineteenth century, laying the foundation of the modern petroleum industry Anticline traps are the most common structural traps in the world. About eighty percent of the world's petroleum can be found in anticline traps. The majorities of anticline traps are produced by sideward pressure, but can also occur from the compacting of sediments. The anticline traps can be filled partially or completely with oil. When it is filled completely with oil, this is known as their spill plane.

Fault Trap

The fault plane must have a sealing effect so that it functions as a fluid migration barrier for reservoir rocks. There are several common types of fault trap:

a) Normal faults — commonly associated with graben (rift) structures.

b) Strike-slip faults — these may not be sealed due to incremental movements, but basement-controlled strike-slip faults commonly produce good anticlinal structures in overlying softer sediments.

c) Thrust faults — commonly associated with compressional tectonics (e.g., the Front Ranges in Alberta).

d) Growth faults — Growth faults typically form in sediments that are deposited rapidly, especially at deltas. Faulting occurs during sedimentation (i.e. syndepositionally), such that the equivalent strata on the downthrow side will be thicker than on the upthrow side.

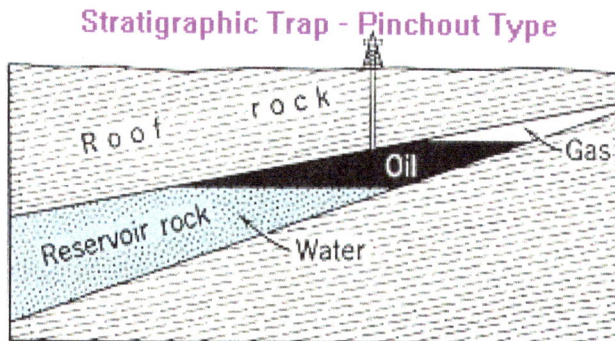

Stratigraphic Trap - Pinchout Type

The throw between corresponding strata declines upwards along the fault plane. Minor fault planes with an opposite throw (antithetic faults) may also form in the strata that curve inward towards the main fault plane. The fault plane is commonly sealed, preventing further upward migration of oil and gas. Fault traps may also form when sandstone beds are offset against the fault plane. Some petroleum traps, however, form in "roll-over" anticlines on the down-faulted block. Growth faults may reduce porewater circulation in sedimentary basins; consequently, undercompacted clays, which may develop into clay diapirs, are often associated with growth faults.

The geometry and timing determine whether faults will be effective in forming fault traps:

1. Dead faults that predate basinal sediments only affect the underlying basement – they play no direct role in hydrocarbon trapping in the younger sedimentary pile.

2. Continuously developing faults (growth faults) — these are active during sedimentation are major petroleum traps (e.g., Niger Delta).

3. Young (late) faults —these form late during sedimentation; depending on their initiation and growth, they may or may not be effective as traps.

4. Late regenerated faults —these are new movements on old faults — they are more likely to destroy than form traps, but may be effective.

Many petroleum fields are closely linked to faulting, but traps that result from faulting alone are less common. There are three common faults – petroleum pool associations:

1. The fault itself makes the trap without an ancillary trapping mechanism such as a fold —normal faults are the most common examples.

2. The fault creates another structure (e.g., a fold or horst) that in turn forms the main trap.

3. The fault may be a consequence of another structure that forms the main trap — e.g., the extensional crestal faults that form above some anticlines.

Salt Dome Trap

Salt dome is a largely subsurface geologic structure that consists of a vertical cylinder of salt (including halite and other evaporites) 1 km (0.6 miles) or more in diameter, embedded in horizontal or inclined strata. In the broadest sense, the term includes both the core of salt and the strata that surround and are "domed" by the core. Similar geologic structures in which salt is the main component are salt pillows and salt walls, which are related genetically to salt domes, and salt anticlines, which are essentially folded rocks pierced by upward migrating salt. Other material, such as gypsum and shale, form the cores of similar geologic structures, and all such structures, including salt domes, are known as diapiric structures. The embedded material in all instances appears to have pierced surrounding rocks. Upward flow is believed to have been caused by the following: gravity forces, in situations where relatively light rocks are overlain by relatively heavy rocks and the light rocks rise like cream to the surface; tectonic (earth-deformation) forces, in situations where mobile material (not necessarily lighter) is literally squeezed by lateral stress through less mobile material; or a combination of both gravity and tectonic forces.

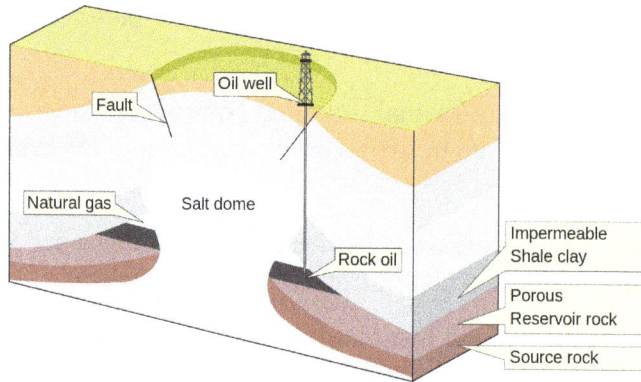

"Classic" salt domes develop directly from bedded salt by gravitational stress alone. Salt domes also may develop from salt walls and salt anticlines, however. In the latter case, the development of the domes results from superposition of gravitational stress on salt masses that initially developed due to tectonic stress.

Physical Characteristics of Salt Domes

A salt dome consists of a core of salt and an envelope of surrounding strata. In some areas, the core may contain "cap rock" and "sheath" in addition to salt.

The size of typical salt domes (including cap rock and sheath) varies considerably. In most cases, the diameter is a kilometre or more and may range up to more than 10 km. The typical salt dome is at least 2 km high (in the subsurface), and some are known to be higher than 10 km.

The cores of salt domes of the North American Gulf Coast consist virtually of pure halite (sodium chloride) with minor amounts of anhydrite (calcium sulfate) and traces of other minerals. Layers of white pure halite are interbedded with layers of black halite and anhydrite. German salt dome cores contain halite, sylvite, and other potash minerals. In Iranian salt domes, halite is mixed with anhydrite and marl (argillaceous limestone) and large blocks of limestone and igneous rock.

The interbedded salt–anhydrite and salt–potash layers are complexly folded; folds are vertical and more complex at the outer edge of the salt. In German domes, when relative age of the internal layers can be deciphered, older material is generally in the center of the salt mass and younger at the edges. Study of halite grains in some Gulf Coast salt domes indicates a complex pattern of orientation that varies both vertically and horizontally in the domes. Mineral grains in the center of a Caspian salt dome are vertical; those at its edge are horizontal.

Shale sheath is a feature that is common to many Gulf Coast salt domes. In shape, it may completely encase the salt (like a sheath), or it may be limited to the lower portions of the salt. It is most common on the deeper portions of salt domes whose tops are near the surface or on deeply buried salt domes. The fluid pressure within the shale is significantly greater than that within the surrounding rocks, and the stratification (bedding planes) of the shale is distorted. Fossils in the shale are older than in surrounding sediments, indicating that the shale came from an older, and therefore deeper, layer.

The strata around salt cores can be affected in three ways: they can be uplifted, they can be lowered, or they can be left unaffected while surrounding strata subside relatively. Uplifted strata have the structural features of domes or anticlines; characteristically they are domed over or around (or both) the core (including cap and sheath if present) and dip down into the surrounding synclines. The domed strata are generally broken by faults that radiate out from the salt on circular domes but that may be more linear on elongate domes or anticlines with one fault or set of faults predominant. Lowered strata develop into synclines, and a circular depression called a rim syncline may encircle or nearly encircle the domal uplift. Unaffected strata develop into highs surrounded by low areas. These highs, called remnant highs or turtleback highs do not have as much vertical relief as the salt domes among which they are interspersed. Present-day structure of strata around salt domes may not in every instance coincide with the present-day position of the salt. This offset relationship suggests that late uplift of the salt dome shifted its center compared with early uplift.

Origin of Salt Domes

In general, salt structures associated with folds have been linked with the same forces that caused the folding. Salt structures in areas without any apparent folding, however, puzzled early geologists and gave rise to a bewildering series of hypotheses. It is now generally agreed that salt structures (and diapiric or piercement structures) develop as the result of gravitational forces, tectonic forces, or some combination of these forces, at the same time or with one force following the other. Whatever the precise circumstance, development of diapiric structures requires a rock that flows.

Although rock flow is difficult to visualize because of slow rates of movement, its results can be clearly seen: stonework that sags, mine and tunnel openings that flow shut, and glaciers of rock salt that move down mountainsides with all the features of glaciers of ice. Given very long periods of time and the relatively high temperature and pressure due to depth of burial, considerable movement of a relatively plastic material such as salt can result. A movement of one millimetre (0.039 inch) a year, for example, over a period of 1,000,000 years would produce a net movement of 1,000 m. The most common rocks that flow are halite, sylvite, gypsum, and high-pressure shale. These rocks also have densities that are lower than consolidated rock such as sandstone, and if buried by sandstone they would be gravitationally unstable. All of them are deposited by normal processes of sedimentation and are widespread in sedimentary strata.

Study of models and natural salt structures have led to a reconstruction of the sequence of events in the development of salt domes First, thick salt is deposited and buried by denser sedimentary strata. The salt and overlying strata become unstable and salt begins to flow from an unreformed bed to a rounded salt pillow. Flow continues into the center of the pillow, doming the overlying strata; at the same time the area from which the salt flowed subsides, forming a rim syncline. The strata overlying the salt, because they are literally spread apart, are subject to tension, and fractures (faults) develop. Eventually, the salt breaks through the center of the domed area, giving rise to a plug-shaped salt mass in the center of domed, upturned, and pierced strata. Upward growth of the salt continues apace with deposition of additional strata, and the salt mass tends to maintain its position at or near the surface. If the salt supply to the growing dome is exhausted during growth, development ceases at whatever stage the dome has reached, and the dome is buried.

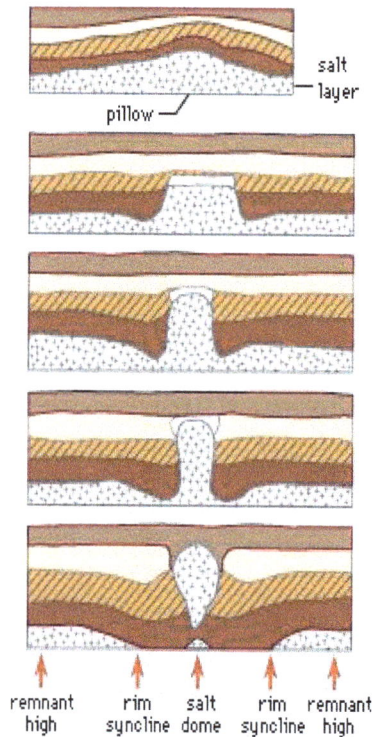

Distribution of Salt Structures

Salt structures develop in any sedimentary basin in which thick salt deposits were later covered with thick sedimentary strata or tectonically deformed or both. With the exception of the shield areas, salt structures are widespread. By their very nature, the classic salt domes generated by gravitational instability alone are limited to areas that have not been subject to significant tectonic stress. Some salt domes do, however, occur in regions that were subject to tectonic stress. Three of the many areas of salt structures in the world are representative of all; these are the Gulf of Mexico region of North America, the North German–North Sea area of Europe, and the Iraq–Iran–Arabian Peninsula of the Middle East.

Economic Significance of Salt Domes

Salt domes make excellent traps for hydrocarbons because surrounding sedimentary strata are domed upward and blocked off. Major accumulations of oil and natural gas are associated with domes in the United States, Mexico, the North Sea, Germany, and Romania. In the Gulf Coastal Plain of Texas and Louisiana, salt domes will be a significant source of hydrocarbons for some years to come. Huge supplies of oil have been found in salt dome areas off the coast of Louisiana. Some individual salt domes in this region are believed to have reserves of more than 500,000,000 barrels of oil. Salt domes in northern Germany have produced oil for many years. Exploration for salt dome oil in the North Sea has extended production offshore.

The cap rock of shallow salt domes in the Gulf Coast contains large quantities of elemental sulfur. Salt domes are major sources of salt and potash in the Gulf Coast and Germany; halite and sylvite are extracted from domes by underground mining and by brine recovery.

Salt domes have also been utilized for underground storage of liquefied propane gas. Storage "bottles" are made by drilling into the salt and then forming a cavity by subsequent solution. Such cavities, because of their impermeability, also have been considered as sites for disposal of radioactive wastes.

Stratigraphic Traps

Stratigraphic traps are hydrocarbon accumulations independent of structural or fault closure. The factors controlling the stratigrahic traps involve facies change, depositional pinch-out, erosional truncation, hydrodynamics, diagenesis, or other factors or a combination.

The prediction of stratigraphic traps relies on a good understanding of complex geological settings. Modern exploration tools such as 3D seismic data generally provide fair imaging of the stratigraphy. Therefore, detailed sequence stratigraphy analysis calibrated by core data and biostratigrahical results must be performed to predict facies variations and geometric architecture of the reservoir-prone section and seal-prone section, respectively, even if the trap has Direct Hydrocarbon Indicator (D.H.I.) support.

The efficiency of bottom seal, lateral seal and top seal is the driving mechanism of the success as any failure in the capacity of the weakest seal will alter hydrocarbon (HC) accumulation.

A series of queries performed on a non-exclusive database provided a statistical evaluation of the relationships between geometric parameters of the traps and HC types. Geometrical aspects, such as the dip of the top of the reservoir and the physical conditions, such as HC density and pressure in bottomhole conditions, control the seal integrity. Porosity and permeability of the reservoir control recovery factors and therefore contribute to economic calculations.

Assessment of stratigraphic traps is best achieved when supported by good understanding of the geology, statistical analysis of geometric parameters and reality check with possible analogues from the database.

Primary Stratigraphic Traps

Primary stratigraphic traps result from variations in facies that developed during sedimentation. These include features such as lenses, pinch-outs.

- Primary pinch out of strata that pinch out updip in less permeable rocks such as shale.

- Fluvial channels of sandstone that are isolated and surrounded by impermeable clay-rich sediments.

- Submarine channels and sandstone turbidites in strata rich in shale.

- Porous reefs that are surrounded by shale.

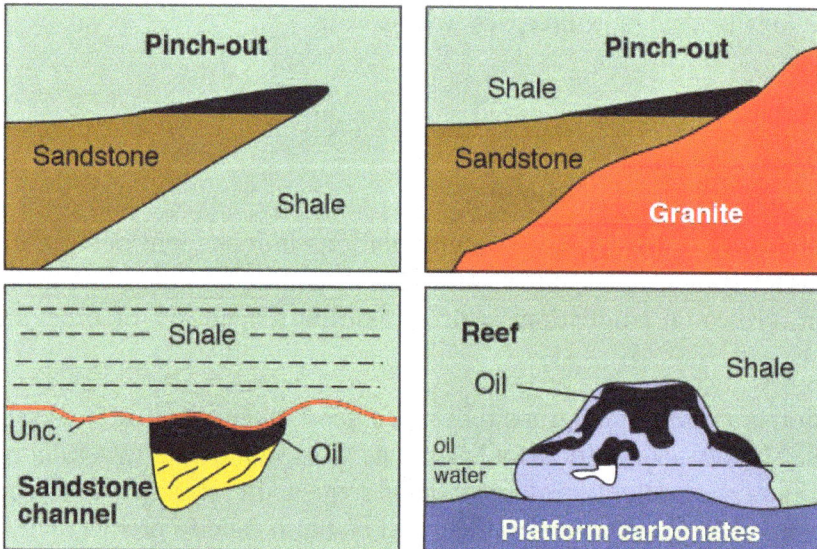

Secondary Stratigraphic Traps

Secondary Stratigraphic Traps results from postdepositional alteration of strata. Such alteration may either create reservoir-quality rocks from non-reservoirs or create seals from former reservoirs. The figure below shows updip porosity loss caused by cementation in previously porous and permeable carbonate rocks. Although the example used is taken from a carbonate setting, similar diagenetic plugging can occur in just about any rock type under the proper circumstances. Porosity occlusion is not limited to only diagenetic mineral cements. Asphalt, permafrost, and gas hydrates are other possible agents that may form seals for this type of stratigraphic trap. Unfortunately, it is often difficult to predict the position of cementation boundaries in the subsurface before drilling, and this type of trap can be a challenging exploration target.

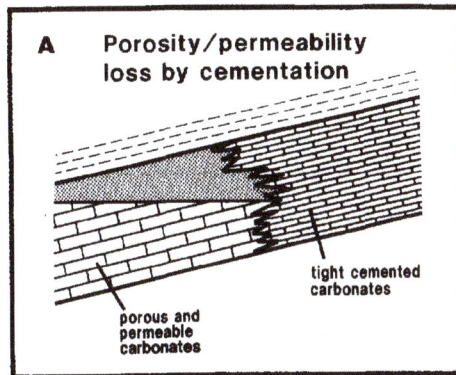

Traps created by postdepositional updip porosity occlusion.

The second type of secondary stratigraphic trap is associated with porosity enhancement that improves reservoir quality in otherwise tight sections. Dolomitization of limited-permeability limestone is a good example Dissolution of framework or matrix material is another porosity- and permeability enhancement mechanism. Porosity enhancement associated with dolomitization and dissolution potentially can create traps on its own. Commonly, though, porosity enhancement is associated with other types of traps as a modifying element. The dolomitized reservoirs of the Scipio-Albion trend in Michigan are a good example of porosity and permeability enhancement along a structural trend.

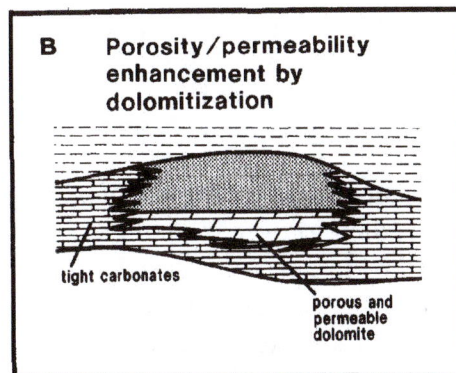

Traps created by postdepositional porosity and permeability enhancement.

Hydrodynamic Traps

In these traps hydrodynamic movement of water is essential to prevent the upward movement of oil or gas. The basic argument is that oil or gas will generally move upward along permeable carrier beds to the earth's surface except where they encounter a permeability barrier, structural or stratigraphic, beneath which they may be tapped.

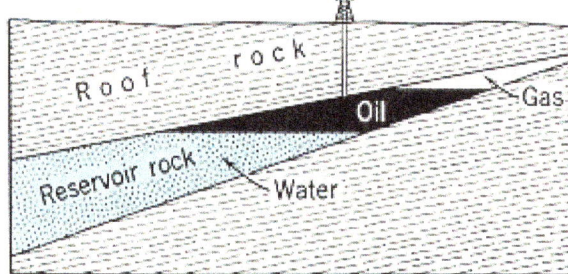

Stratigraphic Trap - Pinchout Type

Where water is moving hydro dynamically down permeable beds, it may encounter upward-moving oil. When the hydrodynamic force of the water is greater than the force due to the buoyancy of the oil droplets, the oil will be restrained from upward movement and will trapped within the bed without any permeability barrier.

Seal/Cap Rock

Caprock is a rock that prevents the flow of a given fluid at a certain temperature and pressure and geochemical conditions.

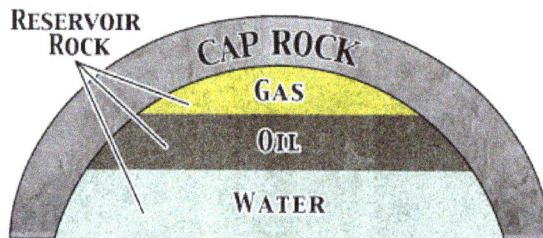

For a long time, the only force causing the movement of oil and gas in the subsurface was believed to be buoyancy. If so, then to form oil and gas accumulation, their migration paths must have been stopped by a roof, i.e., caprock (seal). Clays, shales, carbonates, evaporites, and their combinations can form caprocks. The same rocks react differently to different fluids. In some cases, rocks serve as satisfactory or good conduits for water, but form barriers for oil or gas movement. In some other situations rocks yield oil but stop gas movement, etc. This is determined by capillary forces, the magnitude of which depends on fluid and rock properties (fluid density, fluid viscosity, rock structure, rock wettability) and pore size (capillary forces almost disappear when the pore diameter exceeds 0.5mm). All aforementioned rock and fluid properties are strongly affected by the subsurface temperature and pressure and geochemical environment. Caprock is a rock that prevents the flow of a given fluid at a certain temperature and pressure and geochemical conditions. Therefore, the necessary properties of a rock to act as a seal will be different for different fluids. The same rock with different fluids may or may not have sealing properties up to a complete inversion (caprock - reservoir). The caprocks can be categorized into three types.

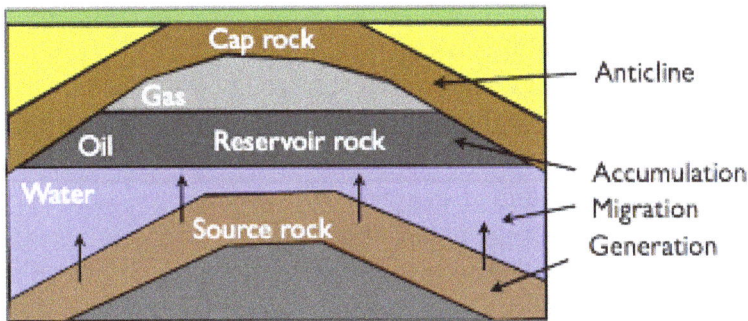

Type I caprocks are typical for argillaceous sequences in a state of continuing compaction; they are developed in areas of young subsidence of Earth's crust, with abnormally high pore water pressure. Sealing properties of these rocks are determined by the amount of capillary pressure at the contact of the reservoir and caprock, the pore pressure of water saturating the caprock, initial pressure gradient of water and the variation of hydraulic forces in the section. Oil and gas accumulations have higher potential energy than that of the formation water. These accumulations can be stable only if this energy is equal to or less than the caprock breakthrough energy. Pore water pressure in compacting argillaceous beds is always greater than the pressure in the adjacent reservoir beds. As a result, sealing capability of the Type I caprocks is determined by hydraulic sealing, by the amount of capillary pressure, and by the pressure at which water begins to flow through caprocks. Just the capillary pressure alone in such caprocks may exceed 100kg/cm^2. This means that the Type I caprocks is capable of confining an oil accumulation having almost any column height. It appears that sealing capability of argillaceous caprocks does not depend on their thickness describes only the aforementioned caprock type.

Type II caprocks are associated with rocks compacted beyond the plasticity limit and having lost ability to swell on contact with water. Such rocks do not contain swelling clay minerals, and interstitial water contains surfactants. Consequently, pore water in these rocks does not have initial pressure gradient. This type of caprocks is encountered mostly in the Paleozoic and Mesozoic sediments of young and old platforms. There are no clear-cut overpressure environments there, but there is a relatively clear hydrodynamic subdivision in the section. The hydrodynamic environment may improve or lower the sealing capability of caprocks. In an extreme case, the water potential in the reservoir may exceed the water potential of the bed overlying the caprock by the value of capillary pressure. In such a situation, the caprock will be open for the vertical flow of hydrocarbons, and the trap will not exist even when potential distribution in the reservoir bed is favorable.

Type III caprocks are typical for rocks with a rigid matrix and intense fracturing. Such caprocks are mainly developed over the old platforms in regions of low tectonic mobility, with no detectable hydrodynamic breakdown of the section. Formation water potential in such regions is practically equal throughout the section and corresponds to the calculated hydrostatic potential.

The correlation between clay mineralogy and their sealing properties are as follows "The permanency in the composition of the silicate layer is a characteristic of the kaolinite group minerals. As a result, replacements within the lattice are very rare and the charges within a layer are compensated. The connection between silicate layers in the C-axis direction is implemented through hydrogen atoms, which prevents the lattice from expanding, ruling out the penetration of water and polar organic liquids. The silicate layer in the montmorillonite mineral group is variable due to a common isomorphic replacement in octahedral and narrower tetrahedral sheets. This replacement results in the disruption of the lattice neutrality. Extra charge that occurs with such replacements is compensated by exchange ions. Ion properties that maintain lattice neutrality in montmorillonite minerals (valence, size of the ion radius, polarization, etc.) define the capability of the lattice to expand along the C-axis. As a result, water and polar organic liquids can penetrate the interlayer spaces. This, in turn, leads to an increase in the volume, which drastically lowers permeability and some other properties, but at the same time improves sealing capabilities. The silicate layer of the illite mineral group is similar to the montmorillonite one. However, the excessive negative charge of the lattice is due mainly to the isomorphic replacements within tetrahedral sheets. The proximity between the source of negative charge and basal surfaces causes a stronger connection between the silicate layers of illite group compared to montmorillonite's."

Admixture of sand and silt degrades the sealing properties of clays. Especially important are the textural changes due to this admixture. Not only the mineral composition of a rock and organic matter content, but also the pore water are important in forming the major sealing properties of clays, such as degree of swelling and compressibility. The relatively low-temperature pore water is retained in argillaceous rocks up to a temperature of 100C to 150C. The temperature of water removal is higher when the concentration of dissolved components is higher. Pore water is located within pores of argillaceous rocks, and at the surfaces and along the edges of individual microblocks and microaggregates that comprise clays. The interlayer water causes swelling in montmorillonites and in degraded illites. The order in water molecules positioning, relative to the clay mineral blocks and aggregates, is rapidly altered with an increase in distance between these blocks and aggregates. Thus, a very important information for the evaluation of the role water plays in the formation of sealing properties is the knowledge of the structural status of the layer in an immediate contact with the particles surface, and the role the cations having different charge density play in the preservation of water molecules structure. Exchange ions play a leading role in the formation of "water clouds" around microaggregates and microblocks of montmorillonite minerals and an insignificant role, with kaolinite minerals. The role played by the illite group minerals occupies an intermediate position. Carbonates caprocks include micro- and fine-grained, massive and laminated limestones. Almost all limestones are dolomitized to some extent and are subject to fracturing. This adversely affects their sealing properties. Carbonates with substantial clay content have laminated texture. As a rule, this results in deterioration rather than an improvement of sealing properties due to the emergence of weakness zones at the contact between different lithologies. Evaporite seals, which are common, include salt, anhydrite, and sometimes shales. It is a common (and probably erroneous) belief that such seals are the best and most reliable. Brittleness of these rocks at the surface conditions contradicts that belief. Besides, cores recovered in the Dnieper-Donets Basin and North Caspian Basin display macro- and microscopic fractures, which sometimes cut monolithic salt crystals. The fractures may be healed by secondary salt, but often contain traces of oil and sometimes gas bubbles. Sometimes core samples are completely saturated with oil. Permeability measured at the surface conditions can reach 100–150mD and even higher. It was established, however, that these rocks easily become plastic even at a relatively low hydrostatic or, even, uniaxial pressure (0100MPa) and the properties change with temperature. Some people considered plasticity as an important sealing property. In this connection, they believe that salt has the best sealing properties. They also believe that the reliability of caprock is not directly related to its thickness. Thus, properties of evaporites as seals change widely during the catagenesis (and in time). Similar changes also affect the other types of seals albeit not so obviously. Inclusions, such as organic matter, silt, clay or carbonate particles degrade sealing properties of evaporites due to the formation of zones of weakness around such inclusions.

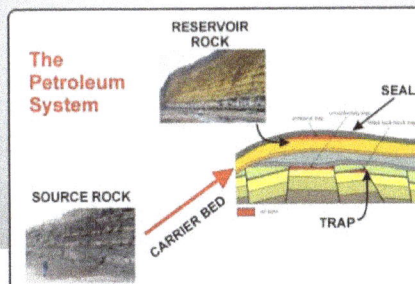

Seals or Cap Rocks
For a trap to have integrity, it must be overlain by an effective seal.
Any rock that is impermeable can act as seal or cap rock but commonly mudstone

References

- Petroleum-traps: geologyin.com, Retrieved 20 May 2018

- Hydrocarbon-traps: geologyin.com, Retrieved 14 July 2018

- Salt-dome, science: britannica.com, Retrieved 05 April 2018

- Stratigraphic-trap: petgeo.weebly.com, Retrieved 23 June 2018

- Types-of-caprocks-in-petroleum-system: geologylearn.blogspot.com, Retrieved 16 March 2018

Chapter 4

Oil and Gas Reservoirs

Oil and gas reservoirs are subsurface pools of hydrocarbons that are contained in fractured or porous rock formations. These are alternatively called as petroleum reservoirs, and can be classified into conventional and unconventional reservoirs. An elaborate study of oil and gas reservoirs has been provided in this chapter, which includes topics that cover oil and gas traps, PVT analysis for oil reservoirs, gas condensate reservoirs, etc.

An oil and gas reservoir is a formation of rock in which oil and natural gas has accumulated within. The oil and gas collect in the small, connected pore spaces of rock and are sealed below ground surface by an impermeable layer of rock. These reservoirs are not "puddles" or "lakes" of oil beneath the surface, as there are no vast open cavities that contain oil.

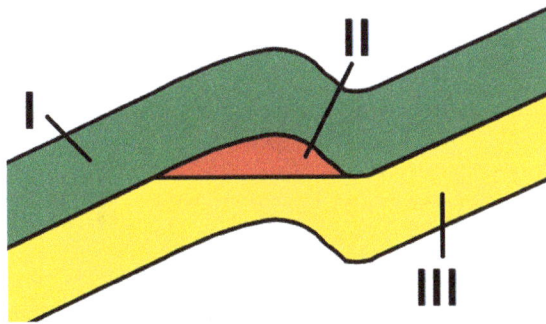

Figure: An oil and gas reservoir. The different components are: I. Seal rock II. Reservoir rock III.

Reservoirs are made of rock known as reservoir rock - shown in Figure above, part II - these are types of rocks that are accessible and contain considerable amounts of oil and gas. In order to be a reservoir rock, it must be both porous and permeable. If both porous and permeable, there are small pockets within the rock where oil or gas can settle and small channels connecting these pockets to allow the oil or gas to flow out of this rock easily when drilled for. These spaces between grains can develop as the formation of rock occurs or afterwards, usually as a result of groundwater passing through the rock and dissolving some of the sediment.

For a reservoir to exist, oil and gas from the source rock (the rock containing kerogen) - shown in Figure above, part III - must migrate into the reservoir rock, which takes millions of years. This migration occurs because oil and gas are less dense than water. This difference in density causes the oil and gas to rise towards the surface so that they are above groundwater with the gas settling above the oil because of their different densities. Migration pathways - a set of well-connected fractures - must exist for this rising to occur.

Finally, there must be some rock preventing the oil and gas from escaping the reservoir rock that they can travel so easily through. For a reservoir to exist the oil and gas must be trapped underground so that they do not produce an oil seep. Traps create the top layer of the reservoir, preventing the fossil fuels from exiting the reservoir rock. There are two main components to a trap: the seal rock and a proper arrangement. The seal rock - shown in Figure above, part I - is the impermeable rock that lies above the reservoir rock, trapping the oil and gas in. Lastly, for a trap to exist there must be the proper arrangement of rocks in the reservoir to create a small, restricted area for the oil and gas to accumulate.

Oil and Gas Traps

All oil and gas deposits are found in structural or stratigraphic traps. You may have heard that oil is found underground in "pools," "lakes," or "rivers." Maybe someone told you there was a "sea" or "ocean" of oil underground. This is all completely wrong, so don't believe everything you hear.

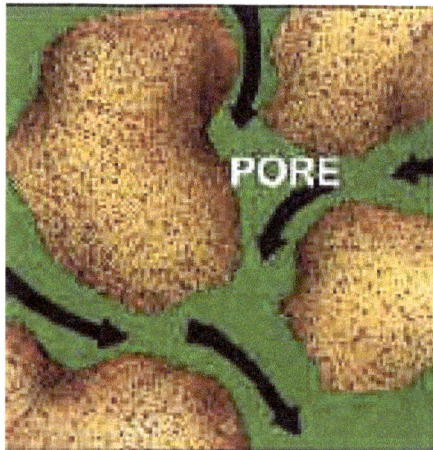

Oil Moving Through Pore Space in Sandstone

Most oil and gas deposits are found in sandstones and coarse-grained limestones. A piece of sandstone or limestone is very much like a hard sponge, full of holes, but not compressible. These holes, or pores, can contain water or oil or gas, and the rock will be saturated with one of the three. The holes are much tinier than sponge holes, but they are still holes, and they are called porosity.

The oil and gas become trapped in these holes, stays there, for millions of years, until petroleum geologists come to find it and extract it.

When you hold a piece of sandstone containing oil in your hand, the rock may look and smell oily, but the oil usually won't run out, and you can't squeeze sandstone like a sponge! The oil is trapped inside the rock's porosity.

Oil Formation and Oil Movement

The very fine-grained shale we talked about previously is one of the most common sedimentary rocks on earth. In many places, thousands upon thousands of feet of shale are stacked up like the pages in a book, deep underground. It is not unusual to have layers in the earth's crust made up mostly of shale that are 4 miles thick. These shales were deposited in quiet ocean waters over millions of years' time.

During much of the earth's history, the land areas we now know as continents were covered with water. This situation allowed tremendous piles of sediment to cover huge areas. The oceans may have left the land we now live on, but the great deposits of shale and sandstone remain deep underground, right under our feet.

The Tiny Gigantic Kingdom

In the deep ocean, far from land, about the only sediment deposited is the fine-grained clastic rock known as shale. But what about the oil and gas? For the answer, we need to move to the ancient oceans that once covered almost all of the earth.

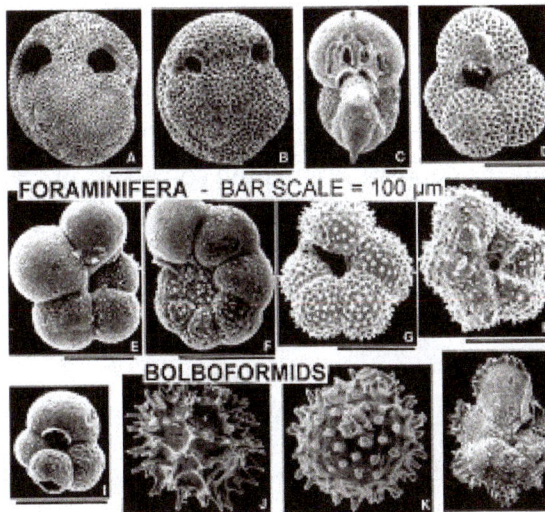

Tiny Microfossils Make Up the Sea-Floor Ooze

A lot of other material is deposited along with the clay or mud-sized sediments. We often think of sharks and whales as being the kings of the deep oceans. Actually, there are other animals that have established giant kingdoms in the sea.the largest and most impressive kingdoms of all! These animals are the various kinds of microscopic creatures both plant and animal. Most of them would fit on the head of a pin. They are tiny, but there are uncountable trillions of them. When these creatures die, they sink to the bottom and become part of the sediments there that will eventually turn into shale.

The animals die by the trillions and rain down on the ocean floor all the time. And since the beginning of life on earth, they have been living their exciting lives in the ocean, dying, sinking to the bottom, and becoming part of the once-living matter that is part of most shale rocks.

It is the trillions of tiny animals that make up most of the gunk (the scientific name for this gunk is "ooze") deposited on the ocean floor. It's a very fine-grained goop containing a lot of organic material mixed with the clay-sized particles that form shale. It is called organic-rich shale.

Later, when thousands of feet of organic-rich shales have piled up over millions of years, and the dead animal bodies are buried very deep (more than two miles down), an amazing thing happens. The heat from deep inside the earth "cooks" the animals, turning their bodies into what we call hydrocarbons, oil and natural gas.

At first, the oil and gas only exist between the shale particles as extremely tiny blobs, left over from the decay of the tiny animals. Then, the Crude Oil Samplesintense pressure of the earth squeezes the oil and gas out of the shale, and the oil and gas fluids gather together in a porous layer and move sideways many miles. On their way, they may meet up with other traveling oil or gas fluids.

Finally, the oil and gas may become "trapped" in a rock formation like sandstone or limestone, a hydrocarbon trap. The oil and gas stay there, under tremendous pressure, until the petroleum geologist comes looking for it. Without a trap, the geologist has no place to drill. All oil and gas deposits are held in some sort of trap.

Crude Oil Samples

The Two Types of Traps

Structural Traps

These traps hold oil and gas because the earth has been bent and deformed in some way. The trap may be a simple dome (or big bump), just a "crease" in the rocks, or it may be a more complex fault trap like the one shown at the right. All pore spaces in the rocks are filled with fluid, water, gas, or oil. Gas, being the lightest, moves to the top. Oil locates right beneath the gas, and water stays lower.

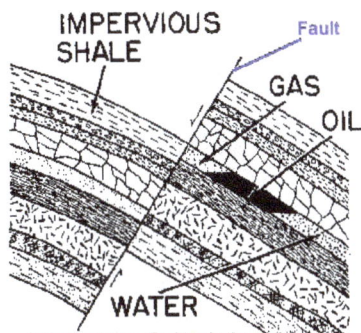

A structural trap. Faulting in the earth has caused vertical movement of the rock layers. Gas and oil cannot pass through the fault boundary, and they are trapped.

Once the oil and gas reach an impenetrable layer, a layer that is very dense or non-permeable, the movement stops. The impenetrable layer is called a "cap rock."

Stratigraphic Traps

Stratigraphic traps are depositional in nature. This means they are formed in place, often by a body of porous sandstone or limestone becoming enclosed in shale. The shale keeps the oil and gas from escaping the trap, as it is generally very difficult for fluids (either oil or gas) to migrate through shales. In essence, this kind of stratigraphic trap is surrounded by "cap rock."

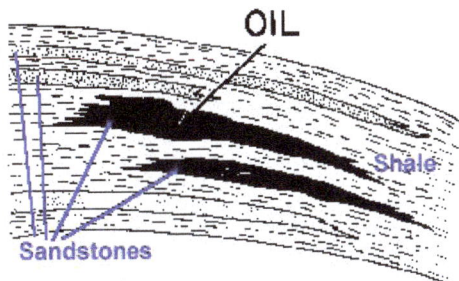

A stratigraphic trap. Oil is trapped in two sandstones which are surrounded by shale. The shale prevents the oil from escaping.

Here are four traps. The anticline is a structural type of trap, as is the fault trap and the salt dome trap.

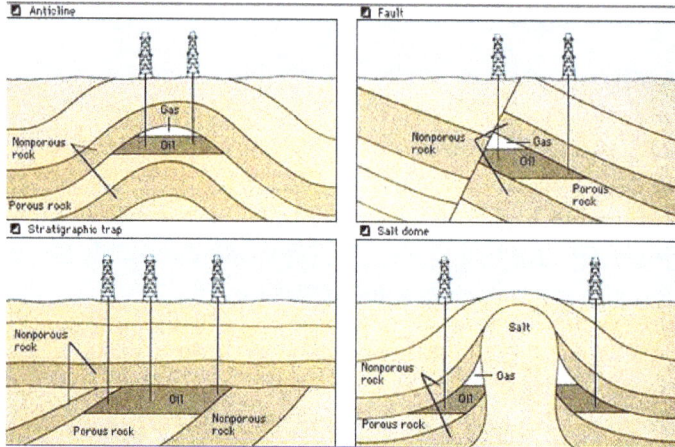

Four Types of Structural and Stratigraphic Traps

The stratigraphic trap shown at the lower left is a cool one. It was formed when rock layers at the bottom were tilted, then eroded flat. Then more layers were formed horizontally on top of the tilted ones. The oil moved up through the tilted porous rock and was trapped underneath the horizontal, nonporous (cap) rocks.

Another Stratigraphic Trap

This hole has been drilled into sandstone that was deposited in a stream bed. This type of sandstone follows a winding path, and can be very hard to hit with a drill bit. The plus is that old stream beds make excellent traps and reservoir rock, and some of these fields are tens of miles long.

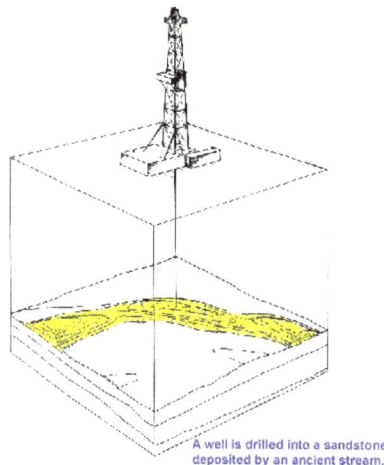

A well is drilled into a sandstone deposited by an ancient stream.

This type of sandstone is usually enclosed in shale, making this a stratigraphic trap.

Just because you drill for oil or gas does not mean that you will find it! Oil and gas reservoirs all have edges. If you drill past the edge, you will miss it. This might explain why your neighbor has a well on his land, and you do not!

Stratigraphic Problems when Drilling

When you drill, you may find a producing reservoir very near the surface. But many other things can happen:

You might drill into a reservoir that has been depleted (all the oil and gas removed) by another well. There may be a new infill reservoir between two wells that could be developed with a third well or one that was incompletely drained. Maybe if you drill a little deeper you might hit a deeper pool reservoir. You might be able to back up and produce a bypassed compartment. The petroleum geologist has to think of all these things when planning a new well.

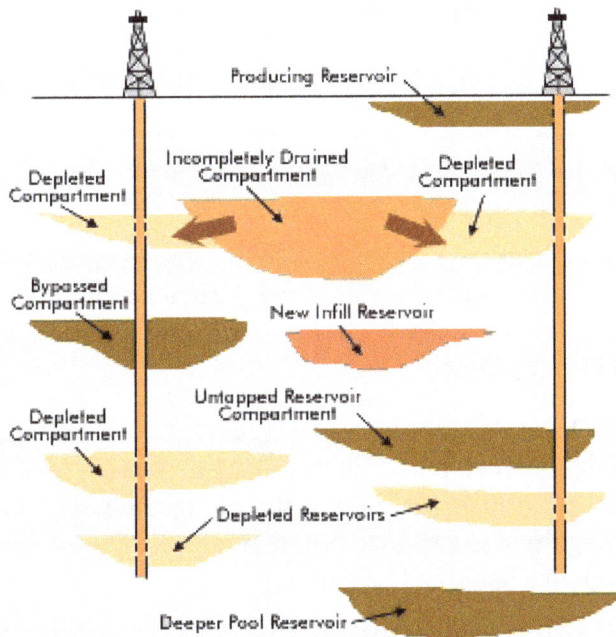

Structural Problems when Drilling

Finally, structures in the earth can give the PG many challenges. Shown in diagram below. Imagine you first drilled the hole on the left into the green layer which represents a nice oil and gas-bearing rock. You have a great well, producing lots of oil and gas.

Then you drilled your second hole to the east (right) of the first one. What happened to that hole?

The oil reservoir has been split in two by the fault, which is nothing but a place in the earth where rock layers break in two. The arrows on the diagram show that the rocks moved DOWN on the LEFT side of the fault and UP on the RIGHT side of the fault. This created a GAP in the oil field right where you drilled your second hole! Incredibly bad luck! Or, bad seismic! Your second hole is a DRY HOLE.

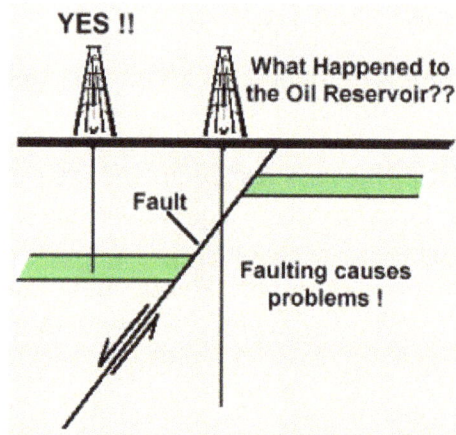

PVT Analysis for Oil Reservoirs

Pressure-Volume-Temperature, PVT, is the study of the physical behavior of oilfield hydrocarbon systems. Information on fluid properties as volumes and phases in the PVT analysis, and how they change with temperature and pressure, is essential to many aspects of Dynamic Flow Analysis.

The need for valid PVT data cannot be over-stressed. Reservoir fluid sampling should be a major objective of any test, particularly during drill-stem testing, as PVT analysis should be performed as early as possible in the life of the field. Once production has started, it may never again be possible to obtain a sample of the original reservoir fluid, which may be continually changing thereafter.

Production analysis engineer, PVT will be used to:

- Assess which phases are present at sandface and at surface.

- Calculate fluid phase equilibrium and phase gravities to correct pressures to datum.

- Calculate fluid viscosity to get from mobility to permeability.

- Calculate pseudopressures and pseudotime to linearize the equations in order to use analytical models.

Phase Equilibrium

Single Component Fluids

Hydrocarbon fluids can exist in two or more separate phases, typically gaseous and liquid, which have different properties. Water may also be present as a separate phase in the reservoir. Reservoir types are classified by their phase behavior, which depends upon the composition, the pressure and temperature. It is the phase behavior that determines the economic recovery in most cases, that makes fluid sampling difficult, and that can sometimes complicate Dynamic Flow Analysis.

The simplest form of phase behavior is for a pure substance, such as methane or water, and ot can be represented on this graphic.

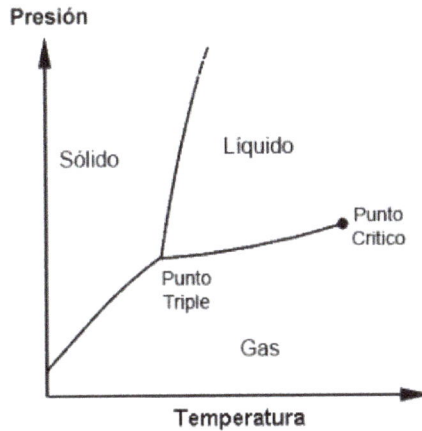

Phase diagram as a product of PVT analysis

The boundary lines between the solid, liquid and gas phases represent values of pressure and temperature at which two phases can exist in equilibrium. There is no upper to the solid-liquid equilibrium line, but the liquid-gas line or vapor pressure curve, terminates at the critical point. At pressures or temperatures above this point only one phase can exist, referred to only as fluid because it has properties similar to both gas and liquid close to the critical point.

Starting in a single-phase liquid state, the volume is increased which causes a sharp pressure reduction due the low liquid compressibility. The point at which the first gas bubble appears is the bubble point. When the volume is further increased the pressure remains the same until the last drop of liquid disappears; this is the dew point. Past the point only gas exists and as the volume increases, the pressure is reduced.

Multi-components Fluids in PVT Analysis

As soon as a mixture of at least two components is considered, phase boundaries become areas rather than lines, due to the combination of the physical properties of two components with different compositions.

Instead of a single vapor pressure curve there are separate lines to represent the bubble points and dew points of the mixture. The two-phase boundary of the system can extend beyond the critical point. With most reservoirs systems it is normal to concentrate only on the liquid-gas equilibrium behaviour.

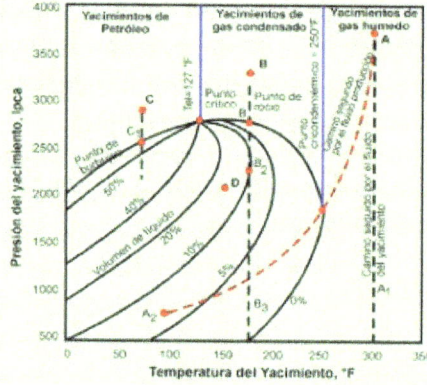

Phase Diagram for a Multi-Component.

Although some hydrocarbons do exhibit solid phases, such as wax precipitation (solid-liquid) and gas hydrate formation (solid-gas). Natural hydrocarbon fluids can contain number of phase loops where liquid and gas phases can exist in equilibrium over a wide range of pressures and temperatures.

When it is above the bubble point, the oil is said to be under-saturated. At the bubble point, or anywhere in the two phase region, the oil is said to be saturated.

Gas Condensate Reservoirs

Gas condensate reservoirs go under two kinds of changes in their lifetime. The phase change and the physical properties change. Both the changes have been handled in this study using the pseudopressure function integral concept. In three phase system accumulating condensate, along with gas phase production, reduces the water production, a very positive impact. Much of the gas phase that goes in the liquid phase is recovered in the form of the liquid. Also it was observed that the mobile liquid condensate cleans the formation. This impact was observed from the continuously decreasing skin factor that was estimated as a function of pressure, an impact never seen before.

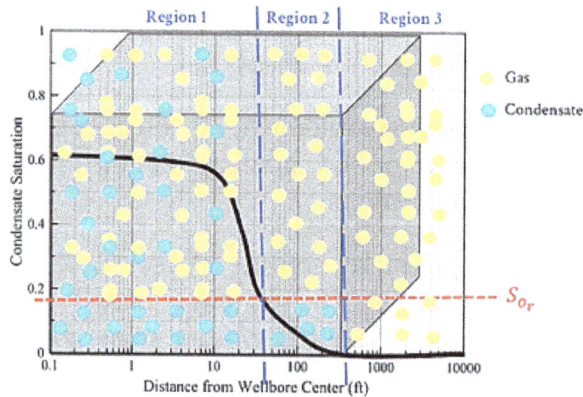

Finally several examples have been solved for two and three phase wells to show the use of the mathematical models developed.

A new method of establishing performance of vertical and horizontal wells completed in gas-condensate reservoirs has been developed. This method does not require relative permeability curves as a function of saturation, instead pressure transient well test data is used to establish the effective permeability as a function of pressure. Surface production data and the pressure transient data are then combined to forecast the well performance. Several new equations of effective permeability in two phase and three phase systems have been introduced from the definitions of producing gas-oil, water-oil, and gas-water ratios for two and three phase systems. Also the new method allows determining the loss in gas well deliverability mathematically after the condensate has begun to liquefy. Thus well efficiency and damage factor can now be calculated analytically. Also well testing equations have been redefined in order to estimate the effective permeability as a function of pressure. Tiab's Direct Synthesis (TDS) technique of pressure transient analysis has also been applied to horizontal gas wells that can also be used for gas-condensate wells.

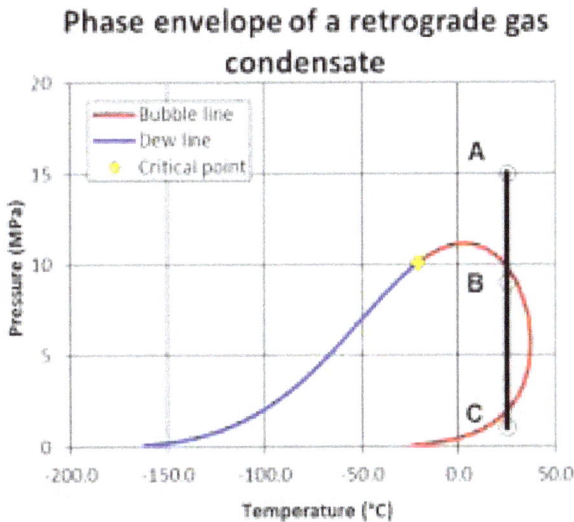

Phase envelope of a retrograde gas condensate

To predict the well performance in multiphase producing environment relative permeability as a function of saturation is used which requires the prior knowledge of the saturation at all the times. Saturation is usually estimated from material balance and reservoir simulation. Also relative permeability curves have to be developed in the laboratory on core samples, an expensive and time consuming process. For individual operators who usually operate on minimum margins of profit, obtaining such data can be an economic challenge. In Oklahoma, regulatory bodies require every well to be tested once a year. Thus a valuable pressure transient data is available on yearly basis. Using that data to forecast well performance can have a profound economic impact on the oil and gas industry. Thus relative permeability curves as a function of saturation have been completely eliminated and it has been shown in this study how to use pressure transient data to develop effective permeability as a function of pressure from the surface measured gas, oil, and water rates and then use it to forecast the well performance.

Phase Behaviour

Gas condensate field produce mostly gas, with some liquid dropout, frequently occurring in the separator. The phase diagram shows the retrograde gas field must have a temperature higher than the critical point temperature. The vertical line on the phase diagram shows the phase changes in the reservoir, while the curve line shows these changes as the fluid cools going up the wellbore and into the separator.

In both cases, liquids drop out as the pressure drops below dew point pressure.

With the retrograde condensate, the %liquid begins to increase to point "A" then decreases with further pressure declines. Thus the name "retrograde" meaning to retreat or go back. So first condensation and then vaporization occurs, and this vaporization can help in further recovery of liquids.

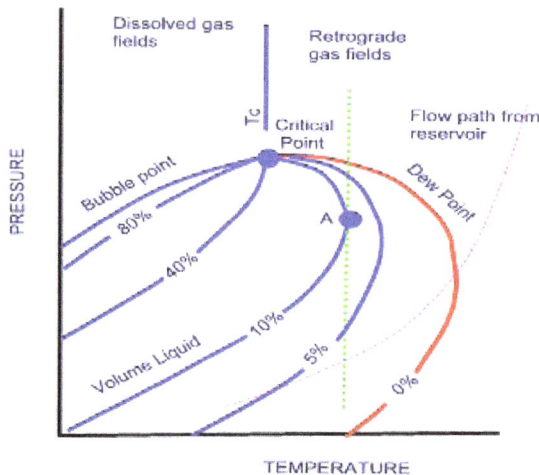

Any hydrocarbon above the dew point line is 100% gas. Any hydrocarbon above the bubble point line is 100% liquid. Hydrocarbons above the bubble point line and close to the critical point are volatile oils. The cricondentherm is the maximum temperature which two phase flow can exist (maximum temperature on the dew point line of the phase diagram). But a field may have both a oil leg and gas cap, which during depletion produces some condensate from the gas.

Fields with active water drive may experience little pressure declines, so condensation occur only at surface and a constant GLR would be expected.

Specific Gravity and Hydrocarbons in Place

GLR=SCF gas to bbl of condensate	$$G=43560 \ A\phi h(1-Swc)/Bg$$
GLR=SCF gas to STB condensate	Can use 43.560 and calculate MSCF to avoid large number.
Specific gravity of produced well fluid Is	Can also calculate G/Ac-Ft
$$\gamma w = \frac{GLR.yg+4585.yo}{GLR+132822.(yo/Mo)}$$	$$G/(A \cdot h)=\frac{43560\phi(1-Swc)}{Bg}$$
$$yo = \frac{141.5}{API+131.5}$$	Gas fraction at surface , fg= $\dfrac{GLR/379.4}{GLR/379.4+350.yo/Mo}$
$$Mo = \frac{6084}{API-5.9}$$	Original gas in place = $fg \cdot G$
Cragoe's empirical relationship, use if Mo is unavailable	Oil in place=OGIP/GLR
	GLR should be in units of SCF/STB

Recovery Estimates

In cases where an active water drive will support pressure, so condensation occurs in the surface separators, the conventional volumetric estimates of gas recovery are possible, where Er=(initial gas volume - final gas volume)/final gas volume. This is the ultimate or displacement recovery. A sweep factor can be added, since so: Er = (Gi-Gf)/Gi*F.

Calculation of Initial Gas and Oil

Calculation of hydrocarbon-in-place volume of a gas-condensate reservoir from the geologic, reservoir, and production data requires a clear understanding of the behavior of oil and gas under various reservoir and surface operating conditions. Because of condensate dropping out from a gas-condensate reservoir fluid due to pressure falling below the dew point, it is necessary to recombine the condensate with the gas in a proper ratio to calculate the original volume of gas-in-place in the reservoir, which would help estimate the potential CO_2 sequestration volume in a condensate reservoir.

The proposed method uses standard charts and simple equations to calculate hydrocarbon-in-place volumes in gas-condensate reservoirs. After the assessment of a basin with known geologic and reservoir data, plots of gas compressibility versus reservoir depth could be prepared for use in resource assessment of gas-condensate reservoirs with little or no data. The proposed method would allow us to see the variability of various geologic and reservoir factors as well as their impact on related parameters, particularly in regions and basins with abnormal pressure and temperature, and therefore would be of great significance in a resource assessment using Monte-Carlo simulation for a probabilistic estimate of resource volumes.

The proposed method, based on correlations established by Rzasa and Katz, provides a means to calculate the gas-in-place volume in a gas-condensate reservoir corresponding to the amount of produced gas and associated condensate. Plots of correlations based on this method are readily available for use. Data required to allow estimates of the gas-inplace volume are:

- Reservoir pressure and temperature, or depth to calculate the required parameters,
- Compositions of oil and gas or their gravities and molecular weights,
- Gravities and production rates of separator condensate and gas,
- Rock porosity,
- Gas or interstitial water saturation, and
- Area and thickness, in the absence of which calculations are based on one acre of reservoir volume.

The following steps summarize the procedure based on the Rzasa and Katz correlations as illustrated by Standing:

1. Calculate average separator gas gravity, if not known, by averaging the gas gravities from various stage separators using gas flow rates from individual stage separators for weighting.

2. Calculate the produced condensate-to-gas ratio by dividing the daily condensate production (in barrels) by the total gas production (in millions of cubic feet).

3. Use the chart showing the condensate-to-gas ratio on the x-axis and the well reservoir-fluid gravity to separator-gas gravity ratio on the y-axis to determine the well reservoir-fluid gravity, based on the known values of condensate gravity and separator-gas gravity (step 1). Mathews, Roland, and Katz worked on Oklahoma City (Wilcox) gases and found a relation that is similar to that for California gases. Standing and Katz studied high-gravity gases, and their relation obviously differs from others.

4. From the correlation between pseudo-critical pressure and temperature and the well-reservoir-fluid gravity, determine the two pseudo-critical parameters for a given reservoir-fluid gravity from step 3.

5. Calculate the pseudo-reduced pressure and temperature, based on known pseudo-critical pressure and temperature (step 4) and the reservoir temperature and pressure.

6. Use pseudo-reduced parameters to determine the gas compressibility factor (Z) using the compressibility factor for natural gases from the study by Standing and Katz.

7. Calculate the hydrocarbon volume of 1 acre-foot of reservoir using porosity and gas (or water) saturation.

8. Calculate the total number of moles in an acre-foot of reservoir rock using the general gas law equation and the known reservoir hydrocarbon pore volume, reservoir pressure and temperature, and gas compressibility factor (Z),

$$N=PV/ZRT$$

Where,

N is pound-moles, P is pressure in psia (pounds per square inch absolute), V is volume in cubic feet, Z is compressibility factor, R is gas constant (10.73), and T is temperature in degrees Rankine (temperature in degrees Fahrenheit + 460).

9. If all the hydrocarbons are to be produced as gas at the surface, the gas volume can be calculated by multiplying the number of moles by 379 (one pound-mole of any gas occupies 379 cubic feet at the standard conditions of 14.7 psia and 60 F).

10. On the other hand, if the condensate is produced with gas at the surface, the following equations can help calculate the number of pound-moles in one barrel of condensate:

Pound-moles in one barrel of condensate = (5.615 × 62.37 × condensate specific gravity)/condensate molecular weight

In the above equation, 5.615 is the cubic feet equivalent of one barrel. The multiple of 62.37 (water density in pounds per cubic feet) and condensate specific gravity gives the density of condensate in pounds per cubic feet. Condensate specific gravity is calculated from its gravity in API (American Petroleum Institute) units by using equation 3:

Condensate specific gravity = 141.5/ (131.5 + condensate gravity in API)

On the right side of equation,

Pound-moles in one barrel of condensate = (5.615 × 62.37 × condensate specific gravity)/condensate molecular weight, the numerator is the condensate mass in pounds in one barrel of condensate, and the denominator is its molecular weight, thereby giving the number of moles.

11. Calculate the number of pound-moles in gas by dividing the gas-condensate ratio (gas volume in cubic feet per barrel of condensate) by 379.

12. Based on known moles of gas and condensate, determine the fraction of produced gas and condensate in the reservoir fluid stream.

13. To obtain produced gas volume per acre-foot of reservoir rock, first multiply the gas mole fraction (from step 12) by total number of moles (from step 8) to get the total number of gas moles in one acre-foot of rock and then multiply by 379 to determine cubic feet of gas.

14. Multiply the condensate mole fraction (from step 12) by total number of moles (from step 8) to get the number of condensate moles in 1 acre-foot of rock and then divide it by number of moles per barrel of condensate (from step 10) to obtain volume in barrels.

Example

Reservoir pressure	3,000 psia
Reservoir temperature	240° F
Porosity	30 percent
Interstitial water saturation	12 percent
Tank condensate production	400 barrels per day
Tank condensate gravity	50° API
Tank gas production	200 MCF per day
Tank gas gravity	1.25 (air = 1)
Primary trap gas production	4,000 MCF per day
Primary trap gas gravity	0.65 (air = 1)

[psia, pounds per square in absolute; MCF, thousand cubic feet; API]

Basis—One acre-foot of reservoir volume

Average separator gas gravity = {(4,000 × 0.65) + (200 × 1.25)}/ (4,000+200) = 0.679
Condensate-gas ratio = 400 × 1,000/ (4,000+200) = 95.2 barrels per million of standard cubic feet (MMSCF)

Using the value of condensate-gas ratio of 95.2, gas gravity of 0.679, and condensate gravity of 50° API, the ratio of reservoir fluid gravity to trap gas gravity is 1.39.

Therefore, the reservoir fluid gravity = 1.39 × 0.679 = 0.944.

Corresponding to gas gravity of 0.944, the pseudo-critical temperature is 435° R and pseudo-critical pressure 647 psia.

Pseudo-reduced temperature = (460+240)/435 = 1.61

Pseudo-reduced pressure = 3,000/647 = 4.64

For the above calculated pseudo-reduced parameters, z = 0.835.

Reservoir hydrocarbon volume in acre-foot rock = 43,560 × 0.3 × (1-0.12) = 11,500 cubic feet.

Using general gas equation, calculate number of moles,

N = PV/zRT = (3,000 × 11,500)/ (0.835 × 10.73 × 700) = 5,500 pound moles.

If all this hydrocarbon were to be produced as gas, gas volume would be = 5,500 × 379 = 2.08 MMCF (million cubic feet) per acre-foot of reservoir rock.

If the reservoir is producing both gas and condensate, we need to find their relative mole fractions.

Basis for calculating moles: One barrel of condensate

The 50° API condensate has a molecular weight of 129 and a specific gravity of 0.78, using the following equation

Condensate specific gravity = 141.5/ (131.5 + condensate gravity in API)

Condensate moles = (5.615 × 62.37 × 0.78)/129 = 273.16/129 = 2.1 per barrel

Gas-condensate ratio = (4,000+200)/400 = 10,500 cubic feet per barrel of condensate
Gas mole per barrel of condensate = 10,500/379 = 27.7

Gas mole fraction = 27.7/ (27.7+2.1) = 0.93

And so, gas-in-place = 0.93 × 5,500 × 379 = 1.9 MMCF per acre-foot of rock

Condensate mole fraction = 2.1/ (27.7+2.1) = 0.07

And so, condensate-in-place = 0.07 × 5,500/2.1 = 183 barrels per acre-foot of rock

Reservoir Gas-in-place = 2.08 MMCF Per Acre of Rock Volume

Some of the parameters (for example the molecular weight) in the above procedure can be calculated from various available equations. One such equation is by Cragoe:

Condensate molecular weight = 6084/ (API-5.9)

There is another equation in Terry's revision of Craft and Hawkins:

Condensate molecular weight = 5954/ (API-8.811)

Therefore, for a 50° API gravity condensate, the molecular weight would be 138 from equation, Condensate molecular weight = 6084/(API-5.9) and 145 from equation, Condensate molecular weight = 5954/(API-8.811).

Similarly, there is correlation between gravity and critical pressure and temperature, through which several researchers have tried to fit the curves and have established equations, but the results vary from one correlation to another.

The correlation between the pseudo-reduced pressure and temperature and compressibility factor (Z) is rather complex. The Standing and Katz chart, which is valid for sweet and dry natural gas, is widely used in the industry because it gives reliable results as long as the percentage of nonhydrocarbon or high-molecular-weight hydrocarbon is low. However, when the percentage of heptane plus (C_7^+) increases, the error increases and dictates the need to use other methods for better results. Of the six methods studied, Elsharkawy and others found the following three methods give better results:

(1) Dranchuk-Abou-Kassem equation,

(2) Dranchuk-Purvis-Robinson method, and

(3) Hall-Yarbrough method.

These methods use iterative solution to calculate the compressibility factor. As an example of these kinds of equations, appendix A presents Dranchuk-Abou-Kassem equations for calculating the values of the gas-compressibility factor (Z).

Average Reservoir Pressure

The average reservoir pressure in a reservoir at a given time is an indication of how much fluid (gas, oil, or water) is remaining in the reservoir. It represents the amount of driving force available to drive the remaining fluid out of the reservoir during a production sequence. When dealing with oil the average reservoir pressure is only calculated when it is undersaturated (flowing pressure above the bubble point). Average reservoir pressure can be estimated in two different ways:

1. By measuring the long-term buildup pressure in a bounded reservoir. The buildup pressure eventually builds up to the average reservoir pressure over a long enough period of time as shown below. Note that this time depends on the reservoir size and permeability (k) (i.e. hydraulic diffusivity).

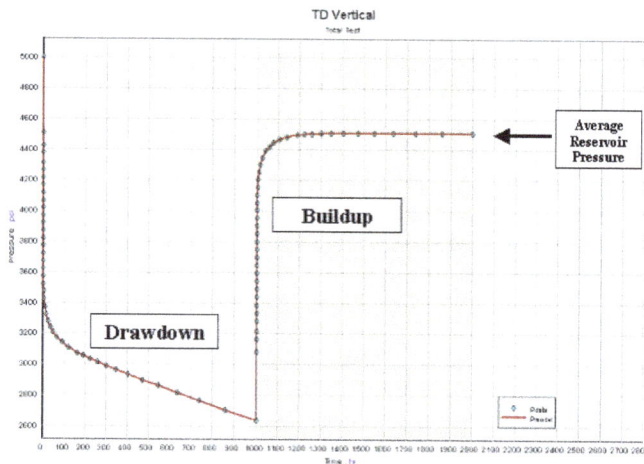

2. Calculating it from the material balance equation (MBE) as described below.

Material Balance Equation (MBE)

Gas

For gas the MBE is defined as the relationship between the original gas in place, initial pressure (pi), cumulative gas production, and the current average reservoir pressure. The basis of the MBE for gas flow is the volumetric balance of all the fluids at a given time. The following equation proposed by Ramagost and Farshad is used to calculate the average reservoir pressure for gas systems. This equation considers that gas is the only mobile phase in the presence of residual fluid saturations (oil and water) in a compressible formation.

$$\overline{p} = \frac{\overline{p}}{Z}\left(1 - c_e\left(p_i - \overline{p}\right)\right) = \frac{p_i}{Z_i}\left(1 - \frac{G_p}{G}\right)$$

$$c_e = \frac{c_f + S_{oi}c_o + S_{wi}c_w}{S_{gi}}$$

Note that this equation is only valid when the term $c_e(p_i - p) < 1$.

Oil and Water

For oil and water the MBE is defined as the relationship between the original fluid in place, initial pressure (pi), cumulative fluid production, total system compressibility (ct), and current average reservoir pressure. The following equation for liquid flow is based on the definition of total compressibility (ct) at a given time. This equation considers the selected fluid (oil or water) as the only mobile phase in the presence of residual fluid saturations, if present, in a compressible formation.

For oil;

$$\overline{p} = p_i - \frac{N_p}{Nc_t}$$

For waters;

$$\overline{p} = p_i - \frac{W_p}{Wc_t}$$

Volumetric Reservoirs

Volumetric reservoirs are defined as completely isolated, closed systems with approximately constant hydrocarbon pore volumes. Volumetric reservoirs are presumed not to gain significant pressure support or fluid influx from outside sources, such as water influx from aquifers or neighboring shale (non-reservoir) layers. On the other hand, nonvolumetric reservoirs exhibit evidence of pressure support or influx of fluids (mostly water) from outside sources, such as aquifers or neighboring shale intervals".

Volumetric or non-volumetric reservoirs can be either oil or gas reservoirs since this classification is intended to describe the characteristics mentioned above, not the type of hydrocarbons being accumulated. In order to determine if a reservoir is o not volumetric, Material Balance Analysis is commonly used: a plot (seldom called "Campbell Plot" for oil reservoirs or "Cole Plot" for gas reservoirs) of the fluids produced at reservoir conditions (also underground withdrawal or "F") versus the sum of the expansive energy terms of the hydrocarbons, water and rock (known as "Et") should shield a horizontal straight line if the reservoir is truly volumetric. If this is the case, then the reservoir is also believed to produce under "Depletion Drive".

Retrograde-Condensate Reservoirs

Retrograde condensation in gas reservoirs occurs when the reservoir pressure drops below the dewpoint of the gas. Condensed liquid will drain downward by gravity when its saturation exceeds the irreducible condensate saturation. Recent studies have shown that in highly permeable, nonfractured water-wet, connate-water-bearing permeable, nonfractured water-wet, connate-water-bearing reservoirs, gravity drainage of condensate can be very effective; the irreducible condensate saturation is low and a large part of the condensate may accumulate at the reservoir bottom within the lifetime of a reservoir. This accumulated condensate can be recovered by properly located production well.

Drainage of Condensate Deposited in the Fractures

Retrograde liquid deposited onto the fracture faces will be imbibed into the matrix by capillarity. Imbibition is counteracted by viscous forces exerted on the liquid by the expanding gas in the matrix. When the resulting imbibition rate is less than the rate of condensation, part of the condensate will flow down the fracture faces to accumulate at the reservoir bottom. When capillary imbibition through the fracture faces is negligible, liquid drainage in a fracture is mathematically analogous to liquid drainage in a porous medium. The basic equations for drainage in a fracture are identical with those for drainage in a porous rock if proper absolute and relative permeabilities are assigned to the fracture. Hence, for negligible capillary imbibition, the recently published analytical model for retrograde condensation and subsequent drainage in a porous rock can be applied unaltered to describe condensate drainage in a fracture. This model describes one-dimensional (1D) vertical drainage of retrograde condensate for a given relation of the reservoir pressure as a function of time.

References

- Oil-and-gas-reservoir: encyclopedia, Retrieved 16 March 2018

- Three-types-of-oil-reservoirs-9892: petropedia.com, Retrieved 27 April 2018

- Oil-and-gas-traps: geomore.com, Retrieved 13 July 2018

- Pvt-analysis-oil-reservoirs: ingenieriadepetroleo.com, Retrieved 17 May 2018

- Gas-condensate, reservoir: petrocenter.com, Retrieved 25 July 2018

Chapter 5

Enhanced Oil Recovery

Enhanced oil recovery (EOR) refers to the implementation of techniques for ensuring the optimal extraction of crude oil from an oil field. Thermal injection, chemical injection and gas injection are the three primary techniques for EOR. This chapter discusses in extensive detail about the processes of enhanced oil recovery that includes primary, secondary and tertiary recovery, infill recovery, etc.

Recovery of Hydrocarbons

The recovery of hydrocarbons is basically a volume displacement process. When a volume of hydrocarbon is removed from the reservoir by production, it will be replaced by a volume of some fluid. Energy is expended in this process. Hydrocarbon recovery mechanisms may be divided into two categories:

 i) Primary Recovery

 ii) Enhanced Recovery

Primary Recovery

Primary recovery is utilization of the natural energy of the reservoir to cause the hydrocarbon to flow into the wellbore. Based on this definition, as long as the hydrocarbon flows into the wellbore, this is primary recovery, even if the hydrocarbon must be artificially lifted to the surface by pumps or some other process. There are many sources of this primary recovery energy of which three are dominant:

 a) Dissolved Gas Drive (Solution Gas Drive)

 b) Gas-Cap Drive

 c) Water Drive

Dissolved Gas Drive

When the reservoir is produced so that gas is permitted to escape from the hydrocarbon liquid in the reservoir, so that two-phase flow (gas and liquid) occurs from the reservoir into the wellbore, the expanding gas will force the oil ahead of the gas into the wellbore.

In order to maximize oil recovery, however, for most reservoirs it is desirable to prevent dissolved gas drive, at least until late in the productive life of the reservoir. As the reservoir approaches depletion, the flowing bottomhole pressures may be reduced to as low a value as possible, in order to recover whatever percentage of remaining hydrocarbons might flow into the wellbore, including solution gas from the oil which will remain in the reservoir (residual oil) at the time the reservoir is abandoned.

Dissolved gas drive can be delayed by injecting water into the water zone beneath the oil, or gas on top of the oil there creating a gas cap, in order to maintain reservoir fluid pressures above the bubble point pressure.

Gas Cap Drive

If a gas cap exists above the oil zone, and wells are drilled and perforated in the oil zone and the bottomhole pressures are sufficiently reduced, the expanding gas cap will force the oil into the wells as the gas interface encroaches into the oil zone. In order for gas-cap drive to exist as a primary recovery mechanism, however, the gas cap must exist naturally.

Water Drive

Most hydrocarbon reservoirs will have a water zone beneath the hydrocarbon. This water is tending to encroach into the oil zone. If wells are drilled and perforated in the oil zone, when the wellbore pressure is reduced, oil flow will be initiated into the well as water encroaches into the oil zone forcing the oil towards the producing wells. If this natural encroachment tendency is to exist, natural energy must be present. There are several possible sources of this natural energy. One source is the expansion of the water as a compressible fluid, as reservoir pressures are reduced. As the reservoir pressure is reduced, the expanding water will push the oil in front of it into the producing wells. Water expansion as a compressed liquid produces more oil than oil as a compressed liquid, not because the compressibility of water is much different to compressibility of oil, but because the total volume of water in the water zone is usually very large when compared to the total volume of oil in the oil zone.

Another source of energy for water drive occurs when the reservoir rock dips upward to the surface where it outcrops. If permeability continuity exists through this rock, as oil is produced from the reservoir, water flows down dip from the surface to replace the oil volume removed. Surface water replenished that water, maintaining a constant hydrostatic pressure on the reservoir fluids.

Secondary Recovery

Secondary recovery is proven technology; indeed, a recent study indicates that 50 percent of all domestic crude oil in the US comes from secondary recovery operations.

Water flooding is inherently more efficient than gas displacement in pressure-maintenance projects and is the preferred process where feasible. Some reservoirs, principally those containing heavy oil that flows only with great difficulty, not only provide poor primary recovery but often are not susceptible to waterflooding. Enhanced oil recovery would be especially useful in some of these reservoirs.

Water Flood

Of the historical techniques used for EOR, water flooding has been the most common. This is not water drive. In water drive, water is encroaching into the oil zone from beneath, but in a true water flood, water is injected down injection wells into the oil zone. Ideally, this creates a vertical flood front, pushing the oil in front of the water toward the producing wells. In a water flood, the water injection wells are placed relative to the oil producing wells in some predetermined pattern based on reservoir characteristics and production history. A common pattern for water flooding for large reservoirs which arc basically horizontal reservoirs is the five spot patterns. This five spot pattern is repeated over the reservoir.

Prior to the initiation of a water flood project for a reservoir, various studies will have been made in designing the water flood. These might include model studies in the laboratory, digital and analog computer simulations, and pilot floods may have been run in a portion of the reservoir as a preliminary study, so that an analysis of the water flood plan might be made.

It is desirable to conduct the water flood so as to maximize the sweep efficiency within economic limits relative to production, so that when the water front from the injection wells breaks into the producing wells, a maximum percent of the reservoir volume will have been swept by the flood. Once this water front reaches the producing wells, further hydrocarbon production will be negligible, in that the wells will now produce essentially water. In order to recover further hydrocarbons, a different EOR technique must now be applied as a tertiary (or third) method for recovery.

Whatever the technique used for enhanced recovery, it is desirable that the mobility ratio of driving fluid be less than the mobility ratio for the driven fluid. The mobility ratio is the ratio of the permeability to the flow of the liquid to the dynamic viscosity of that liquid. The oil ratio mobility ratio will be;

$$[ko/mo] = \text{Oil Mobility Ratio}$$

And, in the case of the water flood, the water mobility ratio of the water will be;

$$[kw/mw] = \text{Water Mobility Ratio}$$

If the mobility ratio of the driving fluid is greater than the mobility ratio of the driven fluid, the driving fluid will tend to channel or finger through the hydrocarbon, tending to bypass the hydrocarbon in the smaller permeability channels, leaving it behind in the reservoir.

Gas Cap Injection

In the gas cap drive injection secondary recovery technique, gas is injected into the gas cap above the oil zone, to pressurize the gas cap. In reservoirs where reservoir fluid pressure is higher than the bubble point pressure, a gas cap may be created by gas injection so that the expending gas cap with further gas injection will displace the oil into the producing wells. As previously discussed, gas cap drive or gas cap drive enhancement is often used as a reservoir pressure maintenance technique.

Enhanced Recovery

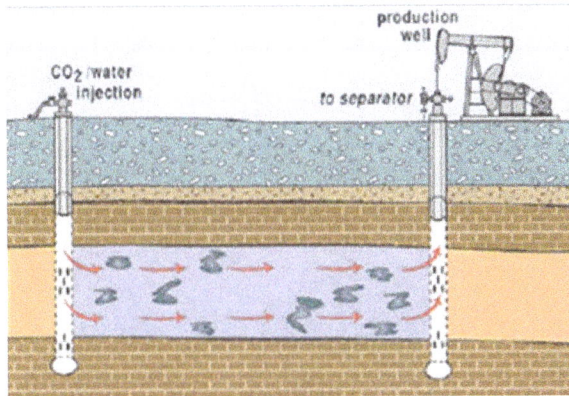

Processes that inject fluids other than natural gas and water to augment a reservoir's ability to produce oil have been designated "improved," "tertiary," and "enhanced" oil recovery processes. The term used in this assessment is enhanced oil recovery (EOR). According to American Petroleum Institute estimates of original oil in place and ultimate recovery, approximately two-thirds of the oil discovered will remain in an average reservoir after primary and secondary production. This inefficiency of oil recovery processes has long been known and the knowledge has stimulated laboratory and field testing of new processes for more than 50 years.

Early experiments with un-conventional fluids to improve oil recovery involved the use of steam and air for combustion to create heat. Current EOR processes may be divided into four categories:

(a) Thermal,

(b) Miscible,

(c) Chemical, and

(d) Other.

Most EOR processes represent essentially untried, high-risk technology. One thermal process has achieved moderately widespread commercialization. The mechanisms of miscible processes are reasonably well understood, but it is still difficult to predict whether they will work and be profitable in any given reservoir. The chemical processes are the most technically complex, but they also could produce the highest recovery efficiencies.

The potential applicability of all EOR processes is limited not only by technological constraints, but by economic, material, and institutional constraints as well.

Thermal Processes

Viscosity, a measure of a liquid's ability to flow, varies widely among crude oils. Some crudes flow like road tar, others as readily as water. High viscosity makes oil difficult to recover with primary or secondary production methods.

The viscosity of most oils dramatically decreases as temperature increases, and the purpose of all thermal oil-recovery processes is therefore to heat the oil to make it flow or make it easier to drive with injected fluids. An injected fluid may be steam or hot water (steam injection), or air (combustion processes).

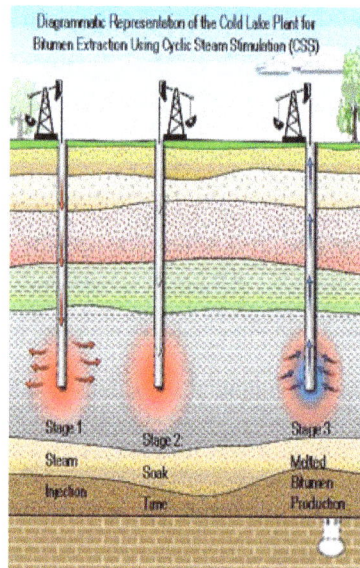

Diagrammatic Representation of the Cold Lake Plant for Bitumen Extraction Using Cyclic Steam Stimulation (CSS)

Steam Injection

Steam injection is the most advanced and most widely used EOR process. It has been successfully used in some reservoirs in California since the mid-1960's. There are two versions of the process: cyclic steam and steam drive. In the first, high-pressure steam or steam and hot water is injected into a well for a period of days or weeks. The injection is stopped and the reservoir is allowed to "soak." After a few days or weeks, the well is allowed to backflow to the surface. Pressure in the producing well is allowed to decrease and some of the water that condensed from steam during injection or that was injected as hot water then vaporizes and drives heated oil toward the producing well.

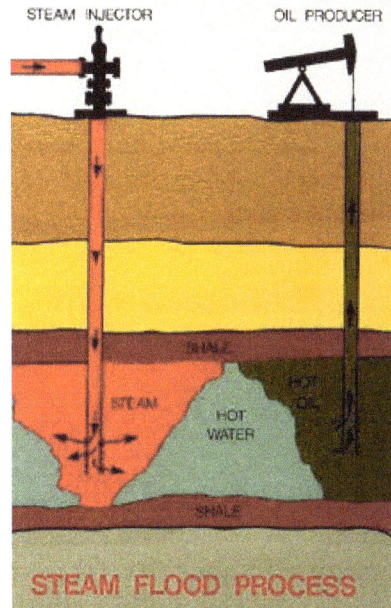

STEAM INJECTOR   OIL PRODUCER

STEAM FLOOD PROCESS

When oil production has declined appreciably, the process is repeated. Because of its cyclic nature, this process is occasionally referred to as the "huff and puff" method. The second method, steam drive or steam flooding, involves continuous injection of steam or steam and hot water in much the same way that water is injected in water flooding. A reservoir or a portion thereof is developed with interlocking patterns of injection and production wells. During this process, a series of zones develop as the fluids move from injection well to producing well. Nearest the injection well is a steam zone, ahead of this is a zone of steam condensate (water), and in front of the condensed water is a band or region of oil being moved by the water. The steam and hot water zone together remove the oil and force it ahead of the water.

Cyclic steam injection is usually attempted in a reservoir before a full-scale steam drive is initiated, partially as a means of determining the technical feasibility of the process for a particular reservoir and partly to improve the efficiency of the subsequent steam drive. A steam drive, where applicable, will recover more oil than cyclic steam injection.

Combustion Processes

Combustion projects are technologically complex, and difficult to predict and control. Injection of hot air will cause ignition of oil within a reservoir. Although some oil is lost by burning, the hot combustion product gases move ahead of the combustion zone to distill oil and push it toward producing wells. Air is injected through one pattern of wells and oil is produced from another interlocking pattern of wells in a manner similar to waterflooding. This process is referred to as fire flooding, in situ (in place) combustion, or forward combustion. Although originally conceived to apply to very viscous crude oils not susceptible to water flooding, the method is theoretically applicable to a relatively wide range of crude oils.

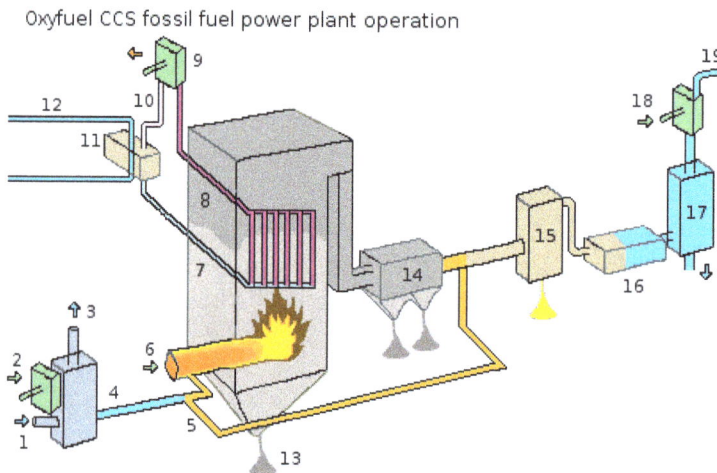

Oxyfuel CCS fossil fuel power plant operation

An important modification of forward combustion is the wet combustion process. Much of the heat generated in forward combustion is left behind the burning front. This heat was used to raise the temperature of the rock to the temperature of the combustion. Some of this heat may be recovered by injection of alternate slugs of water and air. The water is vaporized when it touches the hot formation. The vapor moves through the combustion zone heating the oil ahead of it and assists the production of oil.

Miscible Processes

Miscible processes are those in which an injected fluid dissolves in the oil it contacts, forming a single oil-like liquid that can flow through the reservoir more easily then the original crude. A variety of such processes have been developed using different fluids that can mix with oil, including alcohols, carbon dioxide, petroleum hydrocarbons such as propane or propane-butane mixtures, and petroleum gases rich in ethane, propane, butane, and pentane. The fluid must be carefully selected for each reservoir and type of crude to ensure that the oil and injected fluid will mix. The cost of the injected fluid is quite high in all known processes, and therefore either the process must include a supplementary operation torecover expensive injected fluid, or the injected material must be used sparingly. In this process, a "slug," which varies from 5 to 50 percent of the reservoir volume, is pushed through the reservoir by gas, water (brine), or chemically treated brine to contact and displace the mixture of fluid and oil.

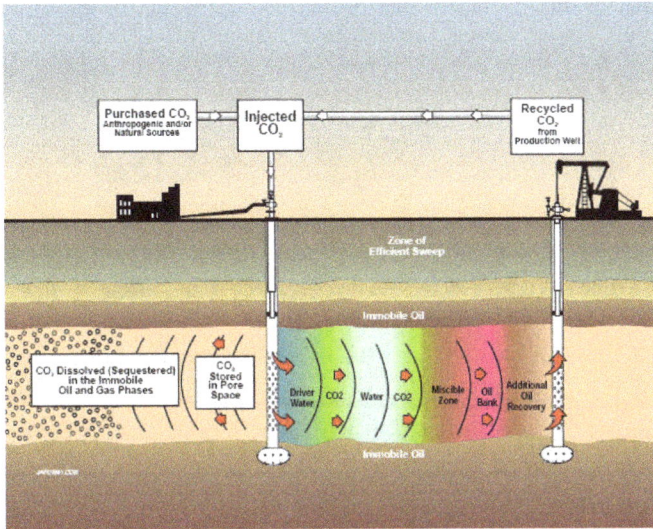

Miscible processes involve only moderately complex technology compared with other EOR processes. Although many miscible fluids have been field tested, much remains to be determined about the proper formulation of various chemical systems to affect complete solubility and to maintain this solubility in the reservoir as the solvent slug is pushed through it. Because of the high value of hydrocarbons and chemicals derived from hydrocarbons, it is generally felt that such materials would not make desirable injection fluids under current or future economic conditions. For this reason, attention has turned to CO_2 as a solvent. Conditions for complete mixing of CO_2 with crude oil depend on reservoir temperature and pressure and on the chemical nature and density of the oil.

Chemical Processes

Three EOR processes involve the use of chemicals: surfactant/polymer, polymer, and alkaline flooding.

Surfactant/Polymer Flooding

Surfactant/polymer flooding, also known as microemulsion flooding or micellar flooding, is the newest and most complex of the EOR processes. While it has a potential for superior oil recovery, few major field tests have been completed or evaluated. Several major tests are now under way to determine its technical and economic feasibility. Surfactant/polymer flooding can be any one of several processes in which detergent-like materials are injected as a slug of fluid to modify the chemical interaction of oil with its surroundings.

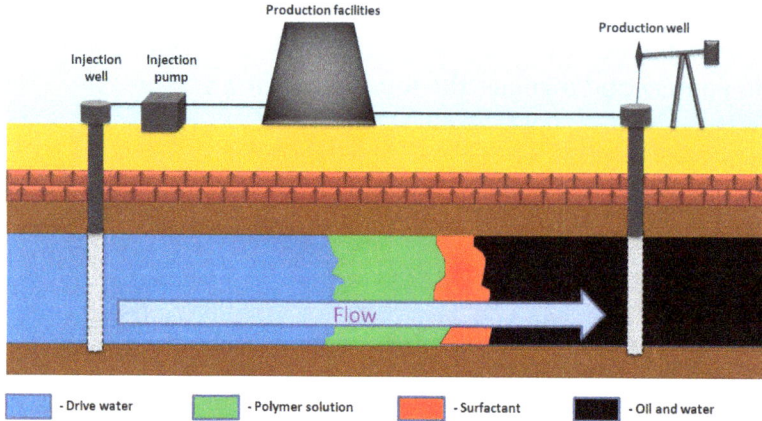

These processes emulsify or otherwise dissolve or partly dissolve the oil within the formation. Because of the cost of such agents, the volume of a slug can represent only a small percentage of the reservoir volume. To preserve the integrity of the slug as it moves through the reservoir, it is pushed by water to which a polymer has been added. The chemical composition of a slug and its size must be carefully selected for each reservoir/ crude oil system. Not all parameters for this design process are well understood.

Polymer Flooding

Polymer flooding is a chemically augmented waterflood in which small concentrations of chemicals, such as polyacrylamides or polysaccharides, are added to injected water to increase the effectiveness of the water in displacing oil.

Alkaline Flooding

Water solutions of certain chemicals such as sodium hydroxide, sodium silicate, and sodium carbonate are strongly alkaline. These solutions will react with constituents present in some crude oils or present at the rock/crude oil interface to form detergent-like materials which reduce the ability of the formation to retain the oil. The few tests which have been reported are technically encouraging, but the technology is not nearly so well developed as those described previously. Reservoirs not considered for alkaline flooding became candidates for other processes.

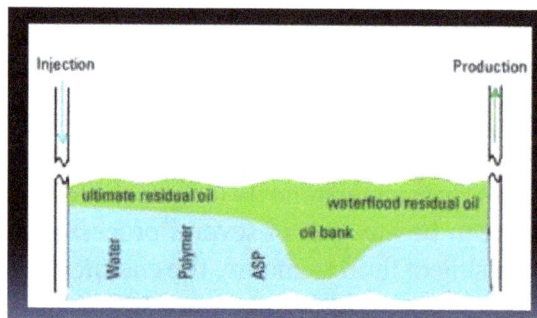

Other EOR Processes

Over the years, many processes for improving oil recovery have been developed, a large number of patents have been issued, and a significant number of processes have been field tested. In evaluating a conceptual process, it should be recognized that a single field test or patent represents but a small step toward commercial use on a scale large enough to influence the supply of crude oil. Some known processes have very limited application, for example, if thin coalbeds lay under an oil reservoir this coal could be ignited, the oil above it would be heated, its viscosity would be reduced, and it would be easier to recover. This relationship between oil and coal is rare, however, and the process is not important to total energy production. Another example involves use of electrical energy to fracture an oil-bearing formation and form a carbon track or band between wells. This band would then be used as a high resistance electrical pathway through which electric current would be applied, causing the "resistor" to heat the formation, reduce oil viscosity, and increase oil recovery. The process was conceived over 25 years ago and has been tested sporadically, but does not appear to have significant potential.

A third process in this category is the use of bacteria for recovery of oil. Several variations have been conceived. These include use of bacteria within a reservoir to generate surface-active (detergent-like) materials that would perform much the same function as a surfactant/polymer flood. Although some bacteria are able to with-stand temperatures and pressure found in oil reservoirs, none have been found that will both successfully generate useful modifying chemicals in sufficient amounts and also tolerate the chemical and thermal environments in most reservoirs. It is uncertain whether nutrients to keep them alive could be provided. Further, any strain of bacteria developed would need to be carefully screened for potential environmental hazards.

Recovery Efficiencies

Experience has determined expected ranges of efficiencies of recovery of hydrocarbons by primary and enhanced techniques. These recovery efficiencies are normally expressed in one of two ways:

  i.)   Percent of Original-Oil-in-Place recovered

  ii.)   Percent of remaining-Oil-in-Place recovered

The ranges of recovery efficiencies for primary recovery and enhanced may be summarized as follows:

  • Primary Recovery Efficiencies

Oil (Percent of Original Oil- in- Place)

         Dissolved Gas Drive   5% to 30%

| Gas-Cap Drive | 20% to 40% |
| Water Drive | 35% to 75% |

- Gas (Percent of Original-Gas-In –Place)

| Gas Expansion and Water Drive | 90% + |

- Enhanced Recovery Efficiencies

Oil (Percent of Original- Oil- In- Place)

Water flood (Secondary Recovery)	30% to 40%
CO_2 Miscible Flood (Tertiary Recovery)	5% to 10%
Steam Drive (Heavy Oil)	50% to 80%

Primary Recovery

Primary recovery is the first stage of petroleum and gas production. Crude oil extraction from a new well relies on the natural rise of the oil due to pressure differences between the oil field and the bottom-hole of the well. Mechanical lift systems such as a rod pump are also a primary recovery method.

Primary recovery is also known as primary production. Primary recovery is less expensive than secondary and enhanced oil recovery (EOR). Enhanced oil recovery techniques are costly and use gases, chemicals, and heat to extract the oil. EOR is expensive and not always useful. Primary recovery takes advantage of the natural tendency for crude oil to rise to the surface once a well punctures the underground oil field.

The crude oil, contained in the ground, is under intense pressure, whereas the hollow well shaft is at lower pressure. Oil will flow rapidly into the area of lowest pressure into the well and up to the surface. When oil under pressure is uncontained, it can result in an oil geyser, spouting from the earth. During primary recovery typically only 5- to 15-percent of a well's total of potential hydrocarbons are extracted.

Hydrocarbons are organic chemical compounds, composed exclusively of hydrogen and carbon atoms. Hydrocarbons can be solids, liquids or gasses. Petroleum and natural gas are made primarily of hydrocarbons. These compounds combust in the presence of sufficient oxygen and produce carbon dioxide, water, and heat. Methane is the primary component of natural gas and is the simplest hydrocarbon because of its structured.

Oil and gas companies will use an estimated ultimate recovery (EUR) calculation to determine if the oil or gas contained in a field has the potential to make the hydrocarbon

Natural Processes Affecting Primary Recovery

Various factors can cause the natural pressures that drive oil to the surface during primary recovery. One common primary recovery method is a gas drive. The gas drive uses the energy of expanding underground gas to force the oil to the surface. Another recovery method is the water drive. Water drives use underground aquifers to exert pressure on the oil. Also, in some shallow and steeply graded oil fields, the oil will drain to the surface with the force of gravity.

As production continues, the reservoir pressure will decrease, and hence the differential pressure will decrease. This decrease in pressure may necessitate the use of an artificial lifting system to continue production. The most common artificial lift for use in primary recovery is the rod pump. The rod pump employs a beam-and-crank assembly to create a reciprocating motion which transfers to vertical lift through a series of plungers and valves. This method is the classic oil derrick with its distinctive bobbing horse head.

Eventually, the primary recovery reaches its limit. This limit happens when the reservoir pressure is too low, or when the mix of gas or water into the output stream is too high. At this point, even artificial lift systems are not economical for the continued extraction of the hydrocarbons.

Solution Gas Drive

Solution gas drive is a mechanism by which dissolved gas in a reservoir will expand and become an energy support to produce reservoir fluid. Solution gas drive has other name, such as dissolved gas drive or depletion drive.

When reservoir pressure is more than the bubble point, no free gas presents in a reservoir and this is called "under saturated reservoir." At this stage, the drive comes from oil and connate water expansion and the compaction of reservoir pore space. Because compressibility of oil and rock is very low, only a small amount of fluid can be produced and typically the volume is around 1-2% of oil in place.

When reservoir pressure reaches a bubble point, oil becomes saturated and free gas will present in a reservoir. The expansion of gas is a main energy to produce reservoir fluid for the solution gas drive. At the beginning, the produced gas oil ratio will be slightly decline because free gas in a reservoir cannot move until it goes over the critical gas saturation. Then gas will begin to flow into a well. In some cases, where vertical permeability is high, gas may migrate up and become a secondary gas cap, which helps oil production.

When pressure gets lower, more gas will be produced and oil production will decline. This will lead to a high producing gas oil ratio. This is not a good sign because reservoir pressure declines sharply with gas production and eventually energy sources in a reservoir will drop and oil cannot be produced. Figure below shows general profiles of reservoir pressure, oil production, and Gas Oil Ratio (GOR) over a period of production.

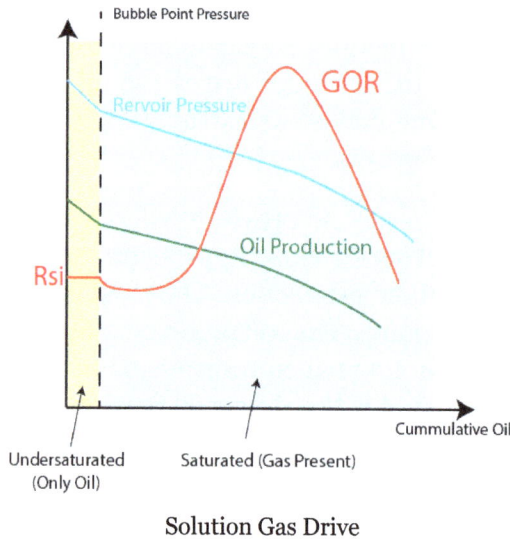

Solution Gas Drive

This is very critical to perform the secondary recovery as water injection to maintain reservoir pressure above the bubble point so as to improve the oil recovery factor. Typical recovery factor from the solution gas drive reservoir is about 5 – 30%.

Gas Cap Drive

Hydrocarbons can experience equilibrium of phases, depending on the pressure and temperature conditions. In many cases, the vapour phase is less dense compared to the liquid, which leads to the accumulation of gas in the upper portions of the porous media forming a gas-cap. In the gas-cap mechanism, the energy comes from its expansion: as oil is produced, the gas-cap expands pushing the gas-oil contact downwards. As gases have high compressibilities, this expansion occurs without significant pressure decrease. Thus, in terms of well positioning, perforations might occur as far as away from the gas cap but not so close to the   water-oil contact to avoid significant water production by coning. The recovery factor from a gas-cap normally will be 20 to 40% – it can be up to 60% in some cases- and the mechanism is dependent upon the following factors:

Size of the gas cap: the degree of pressure maintenance depends upon the gas volume in the gas cap compared to the oil volume.

Vertical permeability: a good vertical permeability will permit the oil moving downward with less bypassing of the gas.

Oil viscosity: an increase of oil viscosity will lead to lower recovery factors, due to the increase in the amount of gas bypassing.

Degree of gas conservation: it is necessary to shut down wells which are producing excessive gas.

Oil Production Rate: lower producing rates will maximize the amount of free gas in the oil zone to migrate to the gas cap which will increase recovery.

Dip angle: steep angle of dip allows better oil drainage.

The recovery factor is higher than for depletion drive mechanisms, given that no gas saturation is formed throughout the reservoir at the same time. Additionally, gas cap reservoirs produce very little or no water.

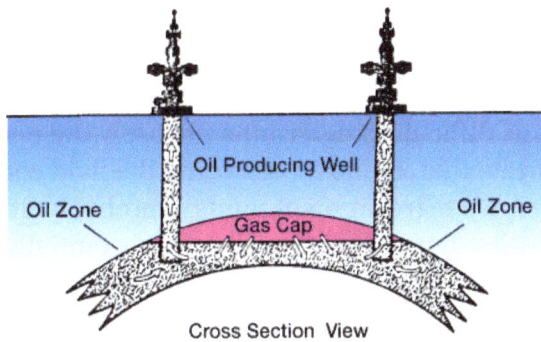

Cross Section View

Water Drive

Water drive is a reservoir-drive mechanism whereby the oil is driven through the reservoir by an active aquifer. As the reservoir depletes, the water moving in from the aquifer below displaces the oil until the aquifer energy is expended or the well eventually produces too much water to be viable.

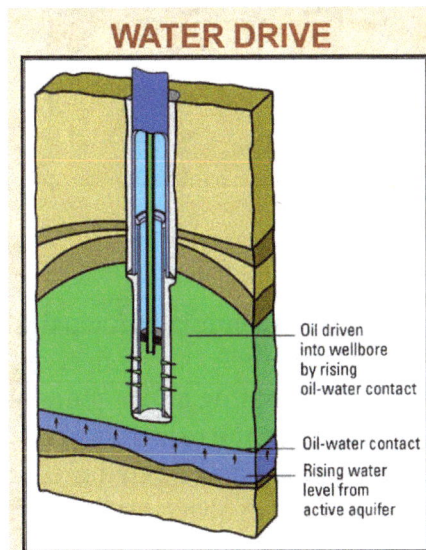

Gravity Drainage

The mechanism of gravity drainage occurs in petroleum reservoirs as a result of differences in densities of the reservoir fluids. The effects of gravitational forces can be simply illustrated by placing a quantity of crude oil and a quantity of water in a jar and agitating the contents. After agitation, the jar is placed at rest, and the denser fluid (normally water) will settle to the bottom of the jar, while the less dense fluid (normally oil) will rest on top of the denser fluid. The fluids have separated as a result of the gravitational forces acting on them. The fluids in petroleum reservoirs have all been subjected to the forces of gravity, as evidenced by the relative positions of the fluids, i.e., gas on top, oil underlying the gas, and water underlying oil. Due to the long periods of time involved in the petroleum accumulation-and-migration process, it is generally assumed that the reservoir fluids are in equilibrium. If the reservoir fluids are in equilibrium, then the gas-oil and oil water contacts should be essentially horizontal. Although it is difficult to determine precisely the reservoir fluid contacts, best available data indicate that, in most reservoirs, the fluid contacts actually are essentially horizontal. Gravity segregation of fluids is probably present to some degree in all petroleum reservoirs, but it may contribute substantially to oil production in some reservoirs.

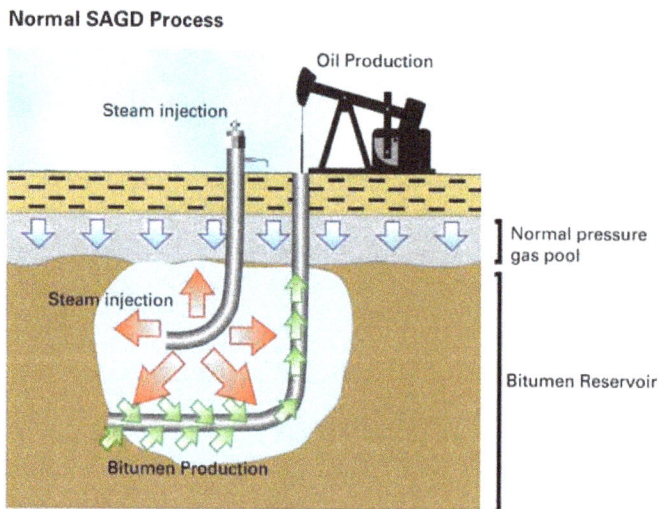

Normal SAGD Process

Reservoir Pressure

Variable rates of pressure decline, depending principally upon the amount of gas conservation. Strictly speaking, where the gas is conserved and reservoir pressure is maintained, the reservoir would be operating under combined gas-cap drive and gravity-drainage mechanisms. Therefore, for the reservoir to be operating solely as a result of gravity drainage, the reservoir would show a rapid pressure decline. This would require the up structure migration of the evolved gas where it later would be produced from structurally high wells, resulting in rapid loss of pressure.

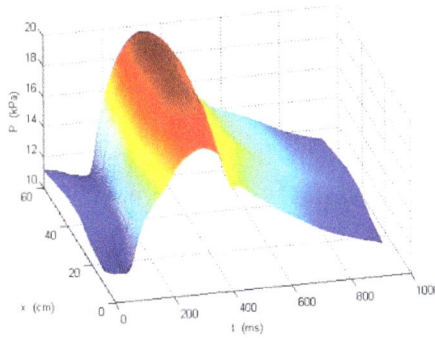

Gas-Oil Ratio

Low gas-oil ratio from structurally low wells. This is caused by migration of the evolved gas up structure due to gravitational segregation of the fluids. On the other hand, the structurally high wells will experience an increasing gas-oil ratio as a result of the up structure migration of the gas released from the crude oil.

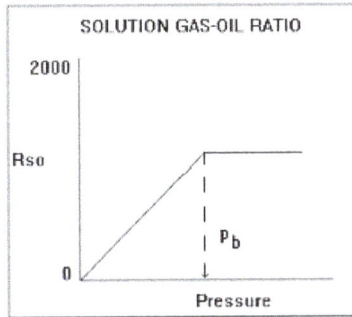

Secondary Gas Cap

Formation of a secondary gas cap in reservoirs that initially were under saturated. Obviously the gravity-drainage mechanism does not become operative until reservoir pressure has declined below the saturation pressure, since above the saturation pressure there will be no free gas in the reservoir.

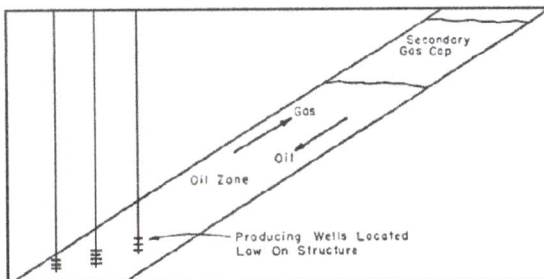

Formation of a Secondary Gas Cap during gas solution liberation

Water Production

Little or no water production. Water production is indicative of a water drive.

Ultimate Oil Recovery

Ultimate recovery from gravity-drainage reservoirs will vary widely, due primarily to the extent of depletion by gravity drainage alone. Where gravity drainage is good, or where producing rates are restricted to take maximum advantage of the gravitational forces, recovery will be high. There are reported cases where recovery from gravity-drainage reservoirs has exceeded 80% of the initial oil in place. In other reservoirs where depletion drive also plays an important role in the oil recovery process, the ultimate recovery will be less.

In operating a gravity-drainage reservoir, it is essential that the oil saturation in the vicinity of the well bore must be maintained as high as possible. There are two basic reasons for this requirement:

- A high oil saturation means a higher oil flow rate
- A high oil saturation means a lower gas flow rate

If the evolved gas migrates up structure instead of toward the well bore, then high oil saturation in the vicinity of the well bore can be maintained. In order to take maximum advantage of the gravity-drainage-producing mechanism, wells should be located as structurally low as possible.

Factors that affect ultimate recovery from gravity-drainage reservoirs are:

- Permeability in the direction of dip
- Dip of the reservoir
- Reservoir producing rates
- Oil viscosity
- Relative permeability characteristics

Combination or Mixed Drive

The driving mechanism most commonly encountered is one in which both water and free gas are available in some degree to displace the oil toward the producing wells. The most common type of drive encountered, therefore, is a combination-drive mechanism as illustrated in Figure. Two combinations of driving forces can be present in combination drive reservoirs. These are:

(1) Depletion drive and a weak water drive and;

(2) Depletion drives with a small gas cap and a weak water drive.

Then, of course, gravity segregation can play an important role in any of the aforementioned drives.

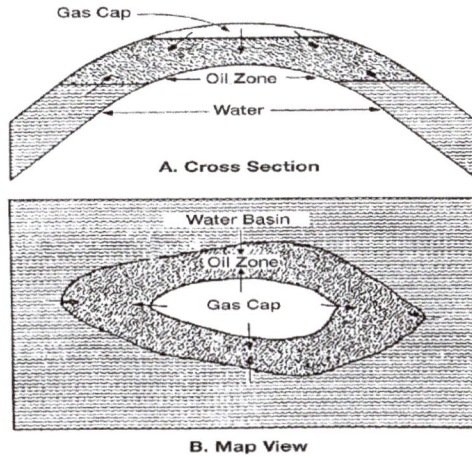

Gas Cap
Oil Zone
Water

A. Cross Section

Water Basin
Oil Zone
Gas Cap

B. Map View

Combination-drive reservoirs can be recognized by the occurrence of a combination of some of the following factors:

a. Relatively rapid pressure decline. Water encroachment and external gas-cap expansion are insufficient to maintain reservoir pressures.

b. Water encroaching slowly into the lower part of the reservoir. Structurally low producing wells will exhibit slowly increasing water producing rates.

c. If a small gas cap is present the structurally high wells will exhibit continually increasing gas-oil ratios, provided the gas cap is expanding. It is possible that the gas cap will shrink due to production of excess free gas, in which case the structurally high wells will exhibit a decreasing gas-oil ratio. This condition should be avoided whenever possible, as large volumes of oil can be lost as a result of a shrinking gas cap.

d. A substantial percentage of the total oil recovery may be due to the depletion-drive mechanism. The gas-oil ratio of structurally low wells will also continue to increase due to evolution of solution gas throughout the reservoir, as pressure is reduced.

e. Ultimate recovery from combination-drive reservoirs is usually greater than recovery from depletion-drive reservoirs but less than recovery from water-drive or gas-cap-drive reservoirs. Actual recovery will depend upon the degree to which it is possible to reduce the magnitude of recovery by depletion drive. In most combination-drive reservoirs, it will be economically feasible to institute some type of pressure maintenance operation, either gas injection, water injection, or both gas and water injection, depending upon the availability of the fluids.

Secondary Recovery

Secondary oil recovery is employed when the pressure inside the well drops to levels that make primary recovery no longer viable. Pressure is the key to collecting oil from the natural underground rock formations in which it forms. When a well is drilled, the pressure inside the formation pushes the oil deposits from the fissures and pores where it collects and into the well where it can be recovered.

But this initial pressure is finite. In order to continue collecting the oil, the pressure must be maintained through other means. These tactics are referred to as secondary recovery techniques.

Methods of Secondary Recovery

There are three main methods of secondary recovery: thermal recovery, gas injection and chemical injection.

The most widely used method of secondary oil recovery is gas injection. Once gas, such as nitrogen or carbon dioxide, is introduced into the reservoir, it expands. This expansion forces oil through the formation and into the well. Gas injection accounts for 60 percent of secondary oil recovery in the U.S.

Thermal recovery is used in about 40 percent of oil wells and relies on heat to facilitate production. Injecting steam or heated water into the reservoir lowers the viscosity of oil, thinning it so that it flows more easily through rock formations and into the well.

The third method, chemical injection, is the least prevalent and is used in less than 1 percent of U.S. oil wells. This approach, also known as chemical flooding, uses solutions composed of micellar polymers and water to reduce the friction between oil and water. Like thermal recovery, this lowers the viscosity of the oil and increases flow.

Secondary recovery techniques are increasingly important in oil and gas exploration since employing these methods can result in a well-being up to 40 percent more productive than with primary methods alone.

Waterflooding

Waterflooding is also known as Water injection. It is a form of this secondary EOR production process.

Used in onshore and offshore developments, water injection involves drilling injection wells into a reservoir and introducing water into that reservoir to encourage oil production. While the injected water helps to increase depleted pressure within the reservoir, it also helps to move the oil in place.

Whether water injection occurs after production has already been depleted or before production from the reservoir has been drained, waterflood sweeps remaining oil through the reservoir to production wells, where it can be recovered.

Water Injection Methods

The water used for water injection is usually some sort of brine, but it can also be made up of other sources that are treated. For example, in some reservoirs water is produced with the hydrocarbons, removed from the production and re-injected into the formation.

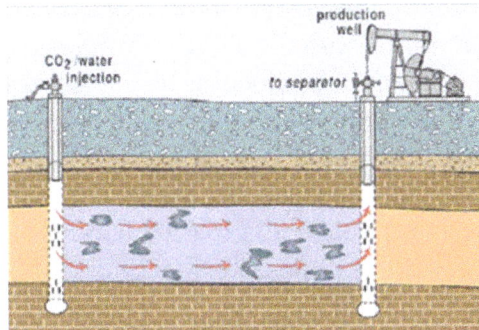

It is important that the water being injected works within the formation. Filtration and processing of the water that will be injected are sometimes necessary to ensure that no materials clog the well pores and that bacteria is not permitted to grow. In an effort to reduce any corrosion within the reservoir, oxygen is often removed from the water, as well.

While production wells can be converted into injection wells, water-injection wells are also drilled specifically for this purpose. Water is then pumped into the reservoir, or gravity can help to push the liquid into the formation. This solution positions water tanks on hills or somewhere above the well, and the water simply is fed into the wellbore.

There are a number of techniques for determining where the water-injection wells should be drilled, as well as established patterns for water-injection wells in relation to production wells. One popular pattern, called the five-spot pattern, involves drilling four water-injection wells in a square around a production well. This is repeated around each production well on the reservoir, resulting in four production wells surrounding each water-injection well, as well.

Other drilling techniques include the seven-spot pattern, which has six water-injection wells surrounding a production well, and the inverted seven-spot pattern, which describes six production wells surrounding one water-injection well.

Also, wells can be drilled in line patterns, rather than spot patterns, where a direct line or staggered line of production wells is followed by a similar line of water-injection wells, and so on. In an edge waterflood, water-injection wells are drilled along the outside borders of the field, and water is injected, with production flowing toward the production wells in the center of the reservoir.

Gasflooding

Gas flooding is also known as miscible flooding. It is one of the leading enhanced oil recovery (EOR) technologies employed for recovering oil that was formerly referred to as either stranded or trapped. Gas flooding is an "enhanced oil recovery" application for injecting miscible (and immiscible) gases into an oil reservoir for increasing oil production.

Gas flooding typically includes CO_2, natural gas or nitrogen as the gas that is injected. Gas flooding tasked place as either a miscible flood or an immiscible flood. Miscible means that the gas that is injected "mixes" with the oil, thereby reducing viscosity and interfacial tension of the oil and rock. Miscible gas flooding also increases oil "swelling" and localized pressure or drive within the reservoir. "Immiscible" flooding means that the gas that is injected into the reservoir does not mix or go into solution. Therefore the purpose of the immiscible flood is to provide the energy or drive by increased pressure. Immiscible flooding does not produce as much oil as miscible gas flooding; however there are certain applications and reservoirs wherein immiscible flooding is well-suited.

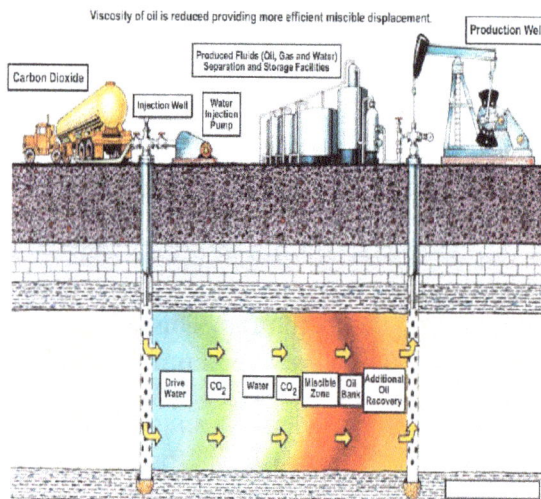

Gas Flooding using CO_2 (CO_2-EOR) as the "miscible" gas for injection.

Tertiary Recovery

Tertiary recovery is the third phase of oil extraction from an oil reserve. This phase of removal allows petroleum companies to remove a significant amount of oil from a reserve which they would not be able to access without these enhanced methods.

There are three primary methods of tertiary recovery:

1. With thermal recovery, the reservoir is heated, often with the introduction of steam. This warms the oil, thinning it so that it loses some of its viscosity and is more apt to flow.

2. In gas injection, the pumping of gases, such as carbon dioxide, nitrogen, or natural gas, into the reservoir is used. The gases expand, and the pressure pushes the remaining oil through the reservoir.

3. Using chemical injection involves injecting polymers, which are long-chained molecules, into the reservoir to lower surface tension and allow the oil to flow more freely. This method is used significantly less frequently than thermal recovery or gas injection.

Thermal EOR

Thermal EOR methods are i) cyclic steam injection (sometimes referred to as Huff & Puff), ii) steamflooding, and iii) Steam- Assisted Gravity Drainage (SAGD).

Cyclic Steam Injection

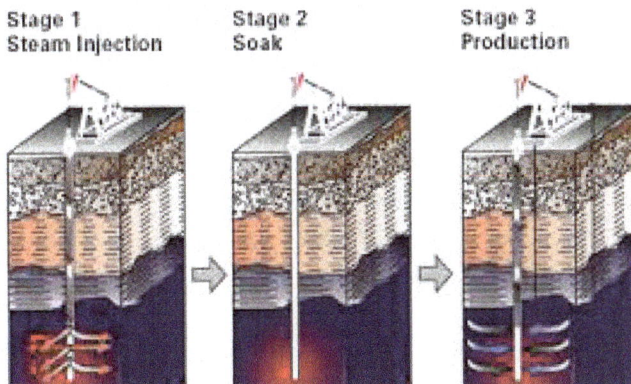

Cyclic Steam Stimulation (CSS)

Stage 1
Steam Injection

Stage 2
Soak

Stage 3
Production

Cyclic steam injection (CSI) is usually done in 3 stages: injection, soaking, and production. First, a predetermined amount of steam is injected into the well to heat the oil in the surrounding reservoir (injection stage). Once the desired amount of steam is injected, the well is shut down to allow the steam to heat reservoir around the well (soaking stage). In the last stage, the injection well is converted to a production mode until the heat is dissipated with the produced fluids (production stage). This cycle is repeated until the response becomes insignificant and economical limits are reached. Obviously, most of the oil is produced in the first few cycles.

CSI is most commonly applied to the reservoirs with thickness greater than 30 ft and depth less than 3000 foot; high reservoir porosity and oil saturation are desirable. Typical recovery factors for this method are in the range of 10-30%. It is very common for wells to produce for a few Huff&Puff cycles before switching to a different thermal EOR method, namely steamflooding.

Steamflooding

For steamflood operations, some wells are used for injection and others are used for production. During the steamflood, high-quality steam is injected into the heavy-oil reservoir; steam heats the oil and pushes it toward a producing well. As opposed to the CSI, during the steamflood, two mechanisms are involved in the recovery process. As in the case of any other thermal method, the first mechanism is viscosity reduction due to increase of temperature; the second mechanism is a physical displacement of oil by steam and hot water. The recovery factors for the steamflood operations are usually higher than one for the CSI and are in the range of 40-60%.

As steamflood matures, a large amount of heat is retained in the reservoir rock and it may become uneconomical to inject steam into the reservoir. In this case the steamflood project is usually stopped or field is converted into a waterflood mode.

Steam-Assisted Gravity Drainage

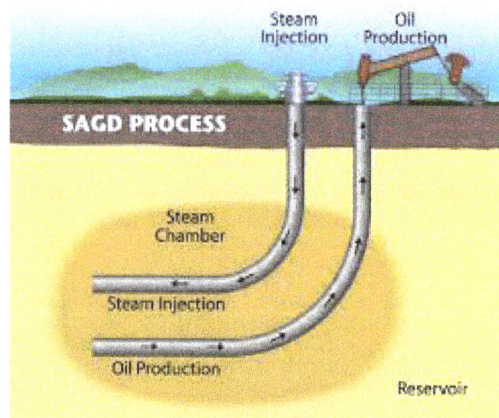

Steam-assisted gravity drainage (SAGD) was first proposed and developed by R. Butler and his coworkers in Imperial Oil. The Foster Creek plant in Alberta Canada was the first commercial SAGD project. Since Foster Creek plant, SAGD has been used increasingly for the recovery of bitumen and heavy oil in Canada. SAGD is considered to be an advanced form of steam injection: here, two horizontal wells are placed, one a few meters above another. The upper well is used as a steam injector; whereas, another well is used as a producer. The steam from the injection well flows towards the perimeter of a steam chamber and condenses. The steam chamber grows both vertically and horizontally heating the surrounding oil with subsequent reduction of oil viscosity. Heated oil and steam condensate flow toward the production well due to gravity.

Method	Location	SOR
Steam Floods	California, USA	~ 4
CSI	California, USA	1.0 - 2.0
CSI	Alberta, Canada	2.0 - 3.3
CSI	Venezuela	~ 0.33
SAGD	Alberta, Canada	2.0 - 3.3

Generally, SAGD leads to a higher recovery factors (about 40-60%) compared to the CSI. These high recoveries are obtained because of the systematic nature of the drainage process. Several different reservoir cutoffs for economical SAGD operations are described by McCorkack. Among the most restrictive cutoffs are: pay thickness greater than 40 ft, permeability greater than 3 Darcy, and absence of top/bottom water. Currently, there is an attempt to improve SAGD technology by introducing noncondensible gas to the steam stream as in SAGP process; or by injecting solvent vapor along with noncondensible gas as in Vapex process. A common feature of the three processes (SAGD, SAGP, and VAPEX) is the use of a pair of long horizontal wells.

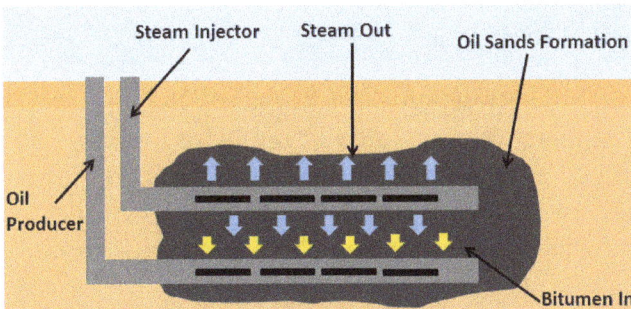

Energy Efficiency

All of the methods described above require steam to be generated on the surface and injected into the petroleum reservoir. Most of the energy consumed during thermal EOR is due to steam generation. The amount of oil produced per amount of steam injected determines the method energy efficiency. For the steam based methods, the commonly used metrics are the volumetric steam-to-oil ratio (SOR) and oil-to-steam ratio (OSR). Since the amount of oil produced and the amount of steam injected vary in time, sometimes it is more meaningful to use cumulative volumes instead of instantaneous. The SOR's are strongly dependent on the quality (density, viscosity, etc) of the produced hydrocarbons, reservoir properties (porosity, permeability), and surface facilities being used. The steam-to-oil ratios for CSI, steamflooding, and SAGD at different locations are presented in Table above.

One of the ways to reduce SOR for a given reservoir is to optimize the design of the surface facilities. For instance, it is beneficial to minimize transportation of hot water to avoid heat losses; or to maximize heat integration between hot and cold process streams to minimize external cooling. On the reservoir side, several optimizations can be done: well placement optimization, injection strategy/production strategy optimization, etc. Commonly, both reservoir and surface facilities optimization are performed during the conceptual stage of the project.

Chemical EOR

Chemical enhanced oil recovery is an advanced technology that addresses the mechanisms that produce oil. The standard way to view oil production is though the oil recovery equation shown below.

$$Np = ED\ EV \cdot OOIP$$

Where:

- Np – Oil Recovery (Production)
- ED – Pore to Pore (Unit) Displacement Efficiency
- EV – Volumetric Sweep Efficiency
- OOIP – Original Oil in Place

Typical oil recoveries are between 20-45% because ED and EV are low. Oil is trapped in the pores due to capillary forces, which accounts for low ED. The capillary force can be minimized by reducing the interfacial tension (IFT), thereby increasing ED. At high IFT the oil drops are stuck in the pores; reducing the IFT results in a more flexible oil drop that is mobile through the reservoir. The figure below shows a simplistic view of how the oil drop is affected by IFT.

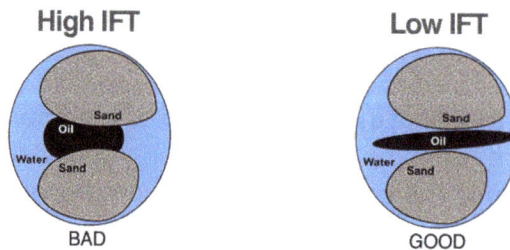

EV is based on water's ability to sweep oil to a producer. The key concept with EV is mobility ratio (M). A simplistic way to understand M is to view it as the ratio of the velocity of water to the velocity of oil. If water moves faster than the oil (high M), it will bypass the oil and break through faster than desired (fingering and channeling affects). If the water moves at the same speed as the oil, it will do a better job sweeping the oil. The figure below shows the effect M has in displacing oil. Polymer agents are added to injection water to viscosify the water and lower its mobility to a more favorable condition.

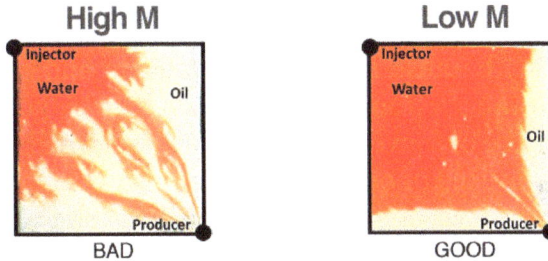

Chemical EOR represents the best methods of improving both efficiencies, ED and EV.

CEOR uses Surfactants and Alkali, and Polymers in an aqueous solution to improve recovery efficiency. Alkalis and Surfactants improve ED, mobilizing the oil that is trapped in the pores. Polymers improve EV by enhancing the water's ability to sweep oil to neighboring production wells by increasing its viscosity.

Each field has a unique oil, water, rock and temperature environment, and therefore the CEOR formulation is specifically developed.

Unlike other forms of EOR, such as CO_2 which requires a nearby source, chemical EOR can be applied anywhere chemical can be transported by road, rail, or boat.

Chemical EOR is implemented in stages and actual size/duration of each stage depends on economics.

Typical chemical flood incremental recovery is in the range of 5-30% OOIP, depending on the process selected and the characteristics of the reservoir.

The figure below shows the stage of implementation for an ASP flood.

1. ASP stage – mobilizes oil trapped by capillary forces creating an oil bank

2. Polymer stage – pushes oil bank to producer

3. Water stage – final push to finish the project

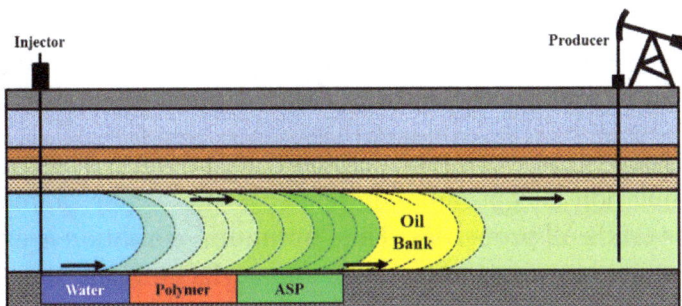

Mobility Control Polymer

Polymer Flooding (P) is the addition of water soluble polymer to viscosify injection water in order to reduce its mobility relative to the oil. Reduction of the mobility attempts to overcome poor sweep efficiency that occurs when viscous oil is displaced with lower viscosity water. Most polymer floods use partially hydrolyzed polyacrylamide polymer (HPAM) because it is widely available and relatively inexpensive. Polymer concentration ranges from 500 ppm to as high as 3000 ppm in reservoirs with highly viscous oil. Use of conventional polymers is limited to ~80°C unless water with very low hardness is used or other chemicals agents are included to stabilize the polymer. However, other types of polymers can push beyond temperature and chemical limitations of HPAM. These include AMPS (copolymer with sulfuric acid) and NVP (with N-vinyl pyrrolidone). Permeability is limited to a minimum of ~25 mD because polymer molecules are large enough to plug the small pore throats found in low permeability formations.

Polymer flooding has been applied successfully throughout the world for over 40 years and it is currently the most widely used chemical flooding technology based on the number of field applications and the mass of polymer injected.

Surfactant Polymer

Surfactant-Polymer (SP) involves adding low concentrations (0.1% to 2%) of a surfactant and co-surfactant and co-solvent and salt to the injection water in order to reduce the oil-water interfacial tension. Polymer is added to increase the solution viscosity in order to overcome viscous instability of low interfacial tension displacement. The SP solution is injected for 20% to 40% pore volume of the target oil-bearing zone, followed by similar volumes of polymer flush.

Conceptually, SP flooding addresses both sweep efficiency and displacement efficiency, limited only to those conditions limiting polymer flooding. Practically, the process is limited to the amount of surfactant and polymer that can be blended together because at some concentration of one in the presence of the other, the solution begins to separate. The concentration of surfactant used must be greater than the critical micelle concentration (CMC) at which point the interfacial tensions become very low for surfactant at "optimum" conditions. The CMC is often quite low. However, both polymer and, to a much greater degree, surfactant are adsorbed by the reservoir rock surface, which decreases their concentration as they progress through the reservoir. Thus the surfactant concentration must be high enough to overcome this adsorption loss.

Designing an SP flood is more complex than polymer flooding. The condition at which a surfactant formulation is "optimum" is related to the water salinity, temperature, and pH, as well as crude oil properties. This "optimum" condition is expressed by very low interfacial tension between the crude oil and surfactant solution. Certain types of surfactant can handle the different reservoir temperature, water salinity, etc. or other unfavorable conditions, but no surfactant is the silver bullet that can handle them all.

Micellar- Polymer

Micellar-Polymer (MP) is similar to SP except higher concentrations of surfactant are used (2% to 12%) and are injected for lower pore volume (5% to 20%). This structuring and many other properties of micellar solutions are sensitive to changes in salinity, temperature, etc. The micellar "slug" is followed by a polymer flush as with SP process.

Micellar-Polymer flooding was widely researched by major oil companies as a means to overcome the limitations of surfactant-polymer flooding – principally the adsorption loss of surfactant. The process proved to be technically very successful in field scale demonstration projects by many major oil companies in the 70's and 80's (10% OOIP to 20% OOIP incremental oil), but was then, and would be now uneconomic due to the high concentrations of surfactant used and their associated costs.

Alkali- Polymer

Alkali-Polymer (AP) flooding involves adding an alkaline agent along with polymer to softened injection water. The water must be softened because the alkaline agents would cause any divalent cations, such as calcium and magnesium, to precipitate, and this solid precipitate will plug most formations. The alkali reacts with components of some oils (saponify) to form "soap" that in the right environment will reduce the interfacial tension sufficiently to overcome capillary forces trapping the oil. The AP solution is injected for 20% to 40% pore volume and followed by a similar volume of polymer flush.

AP flooding has been and is currently being used in full field projects with incremental recoveries of 10% OOIP to 20% OOIP. Of all the chemical flooding processes, it probably has the most limited number of potential applications because not all oils have components that saponify. Also, softening water itself can be complex and costly, depending upon the water hardness and salinity. However, AP flooding is more economic than processes using surfactant because the cost of alkali is much lower than surfactant.

Alkali-Surfactant-Polymer

Alkali-Surfactant-Polymer (ASP) flooding simply adds surfactant (0.05% to 0.5%) to an AP solution to broaden the range of reservoir environment for which the ASP process applies. Many light oils do not contain sufficient amounts of the components that react with alkali to reduce the oil-water interfacial tension sufficiently to overcome capillary forces trapping the oil. Blending surfactant with the alkali overcomes this barrier. As with Alkali-Polymer flooding, the water must have very low divalent ion content (hardness), often requiring softening.

One property of alkaline agents is that they change the reservoir rock chemistry in such a way as to significantly decrease polymer and surfactant adsorption. Therefore, ASP projects use lower concentrations of added surfactant than Surfactant-Polymer or Micellar-Polymer projects. ASP flooding is being applied in field projects around the world with incremental recoveries in some cases greater than 20% OOIP.

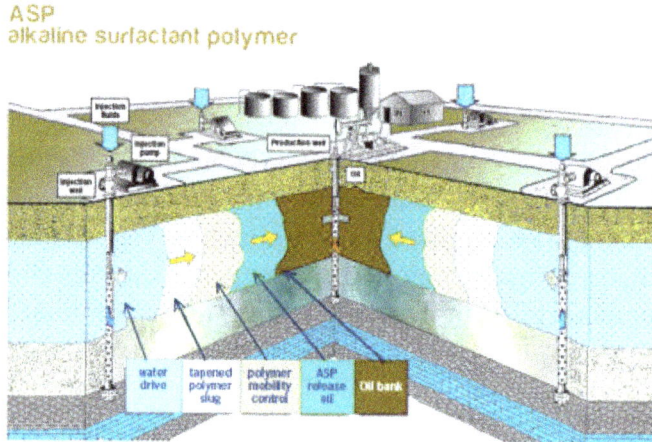

Miscible Gas Flooding

Miscible gas flooding method uses a fluid that is miscible with the oil. Such a fluid has a zero interfacial tension with the oil and can in principal flush out all of the oil remaining in place. In practice a gas is used since gases have high mobilities and can easily enter all the pores in the rock providing the gas is miscible in the oil. Three types of gas are commonly used:

(i) CO_2

(ii) N_2

(iii) Hydrocarbon gases

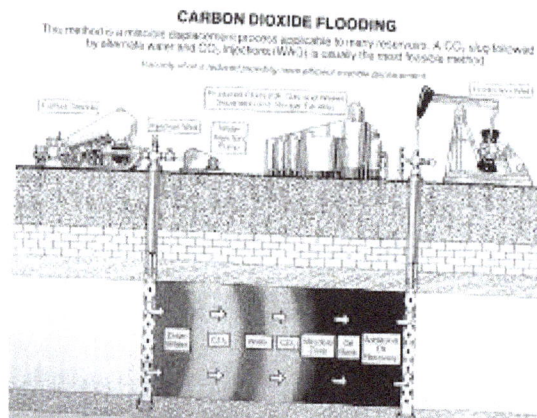

Figure: Miscible WAG Flooding EOR.

All of these are relatively cheap to obtain either from the atmosphere or from evolved reservoir gases. The high mobility of gases can cause a problem in the reservoir flooding process, since gas breakthrough may be early due to fingering, leading to low sweep efficiencies. Effort is then concentrated on trying to improve the sweep efficiency. One such approach is called a miscible WAG (water alternating gas). In this approach water slugs and CO_2 slugs are alternately injected into the reservoir; the idea being that the water slugs will lower the mobility of the CO_2 and lead to a more piston-like displacement with higher flood efficiencies. An additional important advantage of miscible gas-flooding is that the gas dissolves in the oil, and this process reduces the oil viscosity, giving it higher mobilities and easier recovery.

Infill Recovery

Infill drilling is a practice used to speed-up the rate of hydrocarbon recovery by finding the best locations to insert new producers and injectors within the original well pattern. Due to reservoir heterogeneity, wells drilled into a field may not be doing enough to drain hydrocarbons. This can lead to inefficient recovery and sub-optimal reservoir performance.

Infill drilling looks for "sweet spots" within the reservoir that has been by-passed or poorly drained by existing wells, then drills new wells into these spots. This practice is particularly important to reservoirs with low permeability as adding new wells within the original well pattern can help speed up recovery and even add new reserves. In essence, infill drilling reduces the distance between wells for efficient reservoir sweep.

Before Drilling Infill Wells

A successful infill well will boost hydrocarbon production

It is important to understand that infill drilling is applied in mature fields. It is done after the wells in the field have already been producing for a while and a need arises to add more wells. The practice of infill drilling begins with data gathering and data interpretation for proper reservoir description.

This step is crucial in obtaining an accurate understanding of reservoir connectivity and flow barriers. 4D seismic, which gives a graphic view of where reservoir fluids (oil, gas and water) have accumulated with time, can be used to pin-point the right location to drill infill wells.

Infill drilling is not limited to only production wells; infill wells can also be injection wells drilled to sweep by-passed oil zones towards producing wells.

Locating Infill Wells

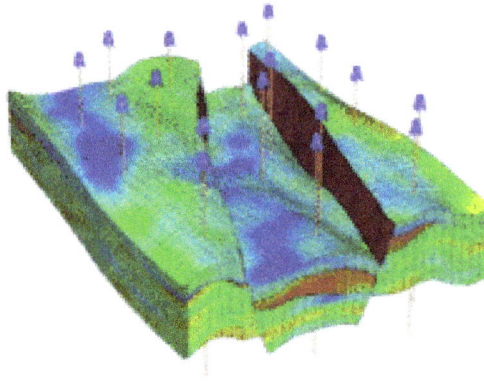

Data gathering through 4D seismic can help in locating sweet spots and by-passed zones.

The choice of "where to drill" an infill well is not as simple as it seems. Multiple possible realistic models should be constructed to determine the optimum location for a new well. This involves testing hundreds or even thousands of potential infill alternatives to know which one will perform best.

Wrongly locating infill wells will result in the new well draining the same area as a previous well thereby eliminating any benefit from the infill well. A poorly located infill injection well may not even be communicating with zones containing hydrocarbons at all leading to a channeling of the injected fluid to wrong zones.

Drilling too many infill wells than necessary is not economical and drilling too few infill wells will result in poor hydrocarbon recovery. Optimum well spacing and location is the goal.

References

- Hydrocarbon-recovery-mechanisms: petroleumandgasengineering.blogspot.com, Retrieved 11 March 2018

- Primary-recovery: investopedia.com, Retrieved 26 May 2018

- Solution-gas-drive-mechanism: drillingformulas.com, Retrieved 13 July 2018

- Drive-mechanisms-in-reservoir: kraken.com.br, Retrieved 26 April 2018

- The-gravity-drainage-drive-mechanism: assignmenthelp.net, Retrieved 30 June 2018

Chapter 6

Petroleum Geology

The study of the origin, accumulation, occurrence and exploration of hydrocarbon fuels is under the domain of petroleum geology. The evaluation of the varied elements of source, trap, reservoir, seal, maturation, timing and migration of sedimentary basins is integral to this field. The topics elaborated in this chapter include oil sands, oil shale, basin and petroleum system modeling, biomarker, etc. are vital for a complete understanding of petroleum geology.

Petroleum Geology is the study of hydrocarbon fuels that deal with the origin, occurrence and exploitation of gas and oil fields. It is basically concerned with some key elements in sedimentary basins such as source, seal, reservoir, trap, timing, etc. A scientist who works in the field of Petroleum Geology is called petroleum geologist. It showcases all aspects of discovery and production. Moreover, with thorough study the geologists are able to locate the exact position of oil deposits and lead.

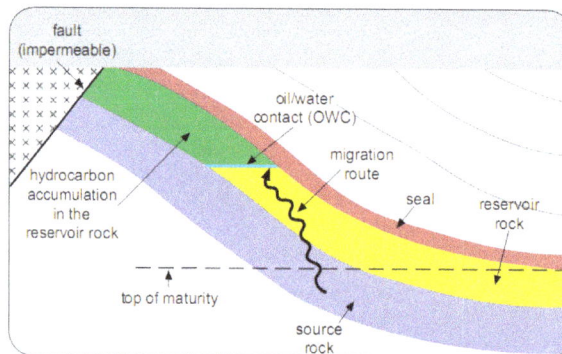

Petroleum Geology plays a vital role in understanding the main geological concepts that affect reservoirs and oil fields. It refers to the certain set geological disciplines that are applied when geologists search for hydrocarbons. The chief disciplines are source rock analysis, exploration stage, basin analysis, production stage, reservoir analysis, and appraisal stage. The analysis helps in determining the permeability and porosity of the drilling samples and also helps to examine the contiguous parts of the reservoir. Petroleum Geology also used to study the sedimentary properties of the rocks.

Elements of Petroleum Geology

Source rocks, trap, seal and reservoir rock are the key elements of petroleum systems

which are provided by the interpretation of data from reflection seismology and electromagnetic geophysical techniques performed in a particular geographic area. Each
of these elements is evaluated in a particular way to determine the potentiality of the
system.

Source Rock

The source rock is a subsurface sedimentary rock units which is made of shale or limestone. It contains the precursors of hydrocarbon formation, organic matters (from
decays of ancient biological species) which were subjected to high temperature for
longtime. The source rock host the processes that involve in the formation oil and
gas until they start to immigrate toward the upper or nearer rock(s) named reservoir
due to the fluidity of oil and gas. The source rock is evaluated using the geochemistry
methods.

Reservoir Rock

This element is a kind of porous or permeable lithological units which retains the im-
migrating oil and gas from source rock. Oil and gas usually accumulate on the top of
water and they are always there relatively to their difference of densities. The reservoir
rock are basically analyzed by means of assessing their porosity a permeability but also
its analysis takes ranges into various fields such as stratigraphy, structural analysis,
sedimentology , paleontology and reservoir engineering disciplines. In case the reservoir has yet been identified, key characteristic crucial to hydrocarbons explorationists
are bulk rock volume and net-to-gross ratio. The bulk rock volume (gross volume of the
rock above the water-hydrocarbons contact) is obtained from of sedimentary packages while the net-to-gross ratio (the proportion of sedimentary packages in a reservoir
rock) estimations are gotten from analogues and wire lines logs. The net volume of
reserves is equal to bulk rock volume multiplied by the net-to-gross ratio.

$$Rv = (BRv)(NtGr).$$

Cap Rock or Seal

It is a lithological units with low permeability which restricts hydrocarbons to escape
from the reservoir. It is made of chalks, shale or evaporites. Its analysis bases on
assessing the extent and thickness to know how much cap rock is efficient to oil and
gas retention. According to lithological deformation that might have been happen,
the cap rock may be found in various types. The tectonic movements the crust experiences cause the anticline and syncline seals and the matter of consequences of their
shapes; the convex form is more enjoyable to petroleum exploration than concave
one. That is why always the seismology experiments are always carried out to assess
how well they can reach the reservoir by aiming at seal with a concave shape as to ease
and make efficient the petroleum exploration.

Trap

The trap is structural or stratigraphic feature that ensures a fixed and firm position of seal and reservoir which avoids the escape of oil and gas.

Maturation

The assessment of the reservoir quality (nature) involves maturation analysis by which they know the length of time of petroleum generation or expulsion.

Migration

Migration is the process of moving oil and gas from the source rock to the reservoir pores when it is trapped after its generation. The main factors of the oil and gas migration are compression, buoyancy, chemical potential; thermal expansion, topography, maturation (increase in volume with time), and gravitational separation of hydrocarbons and water from each other.

Petroleum Systems Identification

With regardless of quantity of oil or gas found in a geographic area, the presence of any quantity shows that petroleum system exists. The identification of petroleum is done through the following steps:

Finding some indication of the presence of petroleum.

Determining the size of petroleum system which is also done through the following steps: first of all the found petroleum occurrences are genetically grouped according to geochemistry characteristics and stratigraphic occurrences. Secondarily, the source which gave rise to the genetically petroleum occurrences is identified using petroleum-source correlation methods. Thirdly, the geographic area the pod of active source found giving rise to all the genetically related petroleum occurrences.

Naming the Petroleum System

The petroleum system nomenclature is essential thing such as the way people have names to identify or differentiate them one from each other. The name of a petroleum system is compound name which has different parts showing the name of the source rock, the part showing the name of the main reservoir rock and another one showing the degree of certainty. The degree of certainty is expressed using signs such as (!) showing that the petroleum is well known, (?) expressing that the petroleum system is speculative while (.) expresses that it is hypothetical. For example the name *Phosphoria-Weber (.)*, *phosphoria* indicates the name of the source rock and *weber* shows the main reservoir rock of the petroleum system which is *hypothetical*. A known system is characterized by having positive petroleum-source rock correlation, hypothetical when that correlation fails indeed with total absence of geochemical evidence.

Petroleum System Mapping

The petroleum system mapping is a way of portraying all the elements of the system basing on geographic (spatial), temporal and stratigraphic extents. The geographic extent of a system is the span of all geographic area in which the active pod source is found. The system size is determined by the geographic extent while the temporal extent traced on the events chart shows the ages of important elements and events such as different process, preservation time, and critical moment of a system in its history. The burial history is explained by the stratigraphic extent which is traced basing on the lithological units of the system.

Practical Applications of Petroleum Systems

Practical application of petroleum system study is optimizing oil and gas exploration, further researches and evaluations of the geographical area in which it is found. Mapping and studying a system helps explorationists to predict if the trap of interest has oil or not. It also helps them locating the most likely accumulations of the petroleum province (region). The best mapping and study of petroleum system is done by linking all elements such as source rock, reservoir rock, seal and overburden rock to the processes of petroleum geology, generation-migration-accumulation.

The system study also allows explorationists to assess risks which might be associated with the entire system, in the petroleum province. Some of those risks are finding new oil accumulations neighboring to the system being explored, that is why in petroleum geology, the oil and gas explorationists, have some risk-associated features such as prospect, play complementary prospect for optimum exploration. A prospect is a potential trap to be drilled to check if it contains sufficient oil for commercial purposes. A play is one or more geological related prospects while a complementary prospect (CP) is a prospect which is dependent on a known system. Hereby other prospects are studied without any knowledge of existing system (being explored). Therefor a better understanding of petroleum system in a province can be explained by the context bellow to associate it with exploration risks.

$$PS_{tot} = PS_{partial} + CP_1 + CP_2 + ... CPn_1 + CPn.$$

PStot means the total system in a petroleum province; CPn means the possible number of unknown oil and gas accumulation that can be discovered by further exploration studies.

A prospect is determined by tree independent assessing variable. The first is a petroleum charge (fluids), the second is the trap that encompasses reservoir, and the entire trap geometry and seal rock. The third is the Timing by which they knew if the trap was formed after or if oil came after. A prospect existence depends on all of the variables above.

P= (v1)(v2)(v3) with assessing variables (v1, v2, v3) which have to be present (1) or absent (0) hence a prospect (P) is absent if one of if one of the variable is absent

Oil Sands

Oil sand: Tar sands specimen close-up photo. From Asphalt Ridge near Vernal, Utah.

Oil sands, also known as "tar sands," are sediments or sedimentary rocks composed of sand, clay minerals, water, and bitumen. The oil is in the form of bitumen, a very heavy liquid or sticky black solid with a low melting temperature. Bitumen typically makes up about 5 to 15% of the deposit.

Process of Oil Removal

The method used to extract bitumen from an oil sand depends upon how deeply the oil sand is buried. If the oil sand is deeply buried, wells must be drilled to extract the bitumen. If the oil sand is close to the surface, it will be mined and hauled to a processing plant for extraction.

Athabasca oil sands mine: Oil sands mining complex along the Athabasca River in Alberta, Canada. The Athabasca Oil Sands are the largest oil sands deposit in the world. It is the second-largest accumulation of oil in the world after Saudi Arabia.

Significance of Oil Sands as a Resource

Most of the world's oil sand resources are located in Alberta, Canada. The Alberta Energy and Utility Board estimates that these contain about 1.6 trillion barrels of oil - about 14% of all of the world's total oil resource. The largest deposit is the Athabasca Oil Sands.

Location of Alberta oil sands deposits: Map showing the location of the Athabasca, Cold Lake, and Peace River oil sands deposits in Alberta, Canada.

Surface Mining

At an oil sands mine, the overburden is stripped away and large mining machines load the sand into trucks that haul it to a nearby processing plant. At the processing plant, the oil sand is crushed and then treated with hot water and chemicals to liberate the bitumen. The liberated bitumen is then separated from the water, blended with lighter hydrocarbons to reduce its viscosity, and pumped through a pipeline to a refinery.

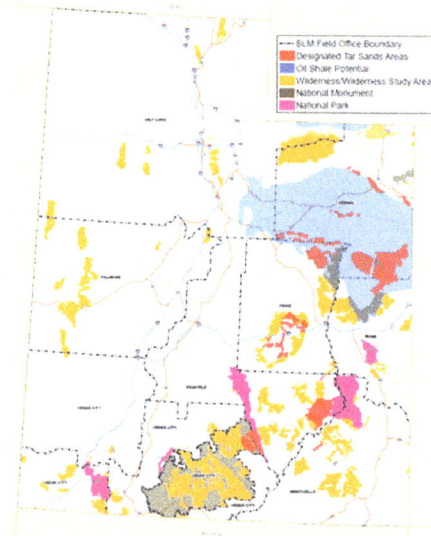

Tar sands areas in Utah: Map showing the location of designated tar sands areas in Utah (red).

Production by Drilling

Bitumen is removed from deeply buried oil sands by drilling wells - a process known as "in-situ recovery." Several wells are drilled down into the oil sand. Then steam and chemicals are pumped down one well. The hot steam and chemicals soften the bitumen, reduce its viscosity, and flush it to extraction wells where it is pumped to the surface. At the surface the bitumen is cleaned, blended with lighter hydrocarbons and pumped through a pipeline to a refinery.

Environmental Concerns

Oil sands mining and processing have a number of environmental impacts. These include: greenhouse gas emissions, land disturbance, destruction of wildlife habitat, and degradation of local water quality. In the United States the water concerns are especially important because the known oil sands and oil shale deposits are located in arid areas of Utah. Several barrels of water are required for each barrel of oil produced.

Oil Shale

Oil shale is any sedimentary rock containing various amounts of solid organic material that yields petroleum products, along with a variety of solid by-products, when subjected to pyrolysis—a treatment that consists of heating the rock to above 300° C (about 575° F) in the absence of oxygen. The liquid oil extracted from oil shale, once it is upgraded, creates a type of synthetic crude oil that is commonly referred to as shale oil. Oil produced from oil shales has potential commercial value in some of the same markets served by conventional crude oil, as it can be refined into products ranging from diesel fuel to gasoline (petrol) to liquefied petroleum gas (LPG). Some of the solid by-products of oil shale processing are unusable wastes, but others have commercial value. These include sulfur, ammonia, alumina, soda ash, and nahcolite (a mineral form of sodium bicarbonate). In addition, spent shale has been used in the production of cement, where the carbon-rich material can enhance the energy balance of the mixture. At the same time, oil shale production has a potentially significant impact on the natural environment, including carbon emission, water consumption, groundwater contamination, and disturbance of land surfaces.

Some confusion has arisen over the terms oil shale and shale oil. Until the early 21st century, those terms respectively referred solely to the organic-rich petroleum source rock described in this topic and to the liquid product obtained from this rock through pyrolysis. In the early 2000s, however, the same terms were applied also to fine-grained impermeable rocks that contain crude oil and to the oil produced from those rocks through hydraulic fracturing.

Formation and Composition of Oil Shales

Geologic Origins

Oil shale was formed from sediments laid down in ancient lakes, seas, and small terrestrial water bodies such as bogs and lagoons. Oil shales deposited in large lake basins, particularly those of tectonic origin, are commonly of considerable thickness in parts. Mineralogically, the deposits are composed of marlstone or argillaceous mudstone, possibly associated with volcanic tuff and evaporite mineral deposits. Major oil shale deposits of this type are the huge Green River Formation (GRF) in the western United States, dating from the Eocene Epoch; oil shales found in the Democratic Republic of the Congo that were laid down in the Triassic Period; and the Albert shale in New Brunswick, Canada, of Mississippian origin.

Oil shale deposited in shallow marine environments is thinner than shale of lacustrine origin but of greater areal extent. The mineral fraction is mostly clay and silica, though carbonates also occur. Extensive deposits of black shales of this variety were formed during the Cambrian Period in northern Europe and Siberia; the Silurian Period in North America; the Permian Period in southern Brazil, Uruguay, and Argentina; the Jurassic Period in western Europe; and the Miocene Epoch of the Neogene Period in Italy, Sicily, and California.

Oil shale deposited in small lakes, bogs, and lagoons is found associated with coal seams. Deposits of this type occur in a sequence found in western Europe dating from the Permian Period and in deposits of northeastern China laid down in the early Cenozoic Era.

Chemical Composition

Oil shales consist of solid organic matter entrained in an inorganic mineral matrix. Chemically, the mineral content consists primarily of silicon, calcium, aluminum, magnesium, iron, sodium, and potassium found in silicate, carbonate, oxide, and sulfide minerals.

The chemical composition of the organic matter is variable. It consists mainly of complex organic molecules containing hydrogen and carbon as well as certain amounts of the heteroatomic elements oxygen, nitrogen, and sulfur. The heteroatomic elements have important effects on the properties of the oil extracted from shales, frequently influencing the choice of upgrading and refining processes, and shales from different regions and different geologic origins are sometimes known for the content of those crucial elements. For instance, the kukersite oil shale of Estonia is noted for being oxygen-rich. Oil shale that originated in saline lake environments, such as the GRF shales of the western United States, tends to be nitrogen-rich, whereas marine oil shales such as those found in Morocco, Egypt, Israel, and Jordan are sulfur-rich.

Mineral Content

The mineral constituents of oil shale vary according to sediment type. Some are true shale in which clay minerals are predominant, such as the Garden Gulch Member of the GRF in Utah. Others, such as the Parachute Creek Member of the GRF in Colorado, are marlstones, containing dolomite or calcite as well as silicate minerals such as clay, quartz, and feldspar.

The various oil shale deposits that have been mined around the world since the early 20th century have ranged from shale to marlstone to carbonate mudstone. All are relatively fine-grained sedimentary rocks, as deposits of coarse sediment such as sand are not compatible with the preservation of organic material. Sandstone found in the Wyoming part of the GRF, for instance, significantly reduces the organic richness of the oil shale.

In the GRF, saline minerals such as nahcolite, trona, and dawsonite, along with a host of other unusual minerals, were most likely formed under extremely saline and stratified conditions in the water of an Eocene lake. The chemical stratification would have created an oxygen-depleted, carbon dioxide-rich environment in the salty bottom layers of the lake, which would have helped to preserve the organic matter, deposit the inorganic minerals, and break down much of the clay carried in as sediment.

Organic Content

The organic matter contained in oil shale is principally kerogen, a solid product of bacterially altered plant and animal remains that is not soluble in traditional petroleum solvents. Kerogen is the source of virtually all crude oil. The richest oil shale ranges from brown to black in colour. Rich oil shale has low density and is flammable, burning with a sooty flame. In addition, oil shale is quite resistant to the oxidizing effects of air. The external structure is commonly laminar; a cross section would show alternating darker and lighter layers, or varves, attributed to annual cycles of organic matter deposition and accumulation. The lamination would have resulted from sedimentation in the quiet waters of a lake or shallow sea, in which either carbonates were precipitated from solution or clay minerals and other silicate minerals were transported as extremely fine detritus.

Some oil shale kerogens are composed almost entirely of identifiable algal remains, whereas other types are a mixture of amorphous organic matter and only some identifiable organic remnants. The main types of algae are Botryococcus, Tasmanites, and Gloeocapsomorpha. Botryococcus is a colonial alga that lives in brackish or fresh water. Permian kerogen from France appears to consist almost exclusively of Botryococcus colonies, as does the kerogen in Carboniferous and Permian torbanites from Scotland, Australia, and South Africa and Holocene coorongites from Australia. Tasmanites is a marine alga the remains of which make up nearly all the kerogen of tasmanites in Australia (Permian) and Alaska (Jurassic-Cretaceous). The remains of Tasmanites also are present in the Lower Toarcian shales of the Paris Basin in France and the Lower Silurian shales of Algeria. Gloeocapsomorpha prisca is a marine alga that makes up the kerogen found in the kukersite oil shales of Estonia and adjacent Russia. Oil shale in Queensland, Australia, contains kerogen derived from planktonic lacustrine algae.

Commonly, only a minor part of the kerogen in oil shale is made of recognizable organic remnants. The rest is amorphous, probably because of alteration by microbes during sedimentation. Amorphous organic material (known as sapropelic matter) is found in thick accumulations in the Permian Irati shale of Brazil and in the Eocene GRF. The organic material may have been derived from planktonic organisms (e.g., algae, copepods, and ostracods) and from microorganisms that lived in the sediment (e.g., bacteria and algae).

World Oil Shale Resources

Oil shale is found in more than 30 countries around the world, yet, on a global scale, its development has been economically attractive for only a few brief periods since the early 20th century. Only in a few locations, where specific conditions have made its exploitation feasible, has oil shale been developed for any considerable period of time. Developed oil shale formations include the kukersite deposits of northern Estonia (extending into northwest Russia), the Fushun deposits of northeast China, and the Irati Formation of southern Brazil. In addition, the large and rich Green River Formation (GRF) in the western United States has attracted commercial interest periodically, depending on the price of conventional crude oil.

Table: Shale oil resources and production of the world.

leading countries	in-place resources		production (2008)	
	million barrels	million metric tons	thousand barrels/day	thousand metric tons/year
United States	4,291,363	617,956		
China	354,430	47,600	13	641
Israel	250,000	36,000		
Russia	247,883	35,470		
Jordan	102,000	14,688		
Democratic Republic of the Congo	100,000	14,310		
Brazil	82,000	11,734	3.8	200
Italy	73,000	10,446		
Morocco	53,381	8,167		
Australia	31,729	4,531		
Estonia	16,286	2,494	9	507
Canada	15,241	2,192		
world total	5,684,497	815,198	25.8	1,348

The estimated shale oil resources of the top 12 countries together account for 99 percent of the world's total reserves and production. The "in-place resources" of each country are commonly given in oil equivalents—that is, the number of barrels or metric tons of oil that are estimated to be contained in the oil shale deposits located in that country. In turn, the estimated values for contained oil are based primarily upon the amount of oil extracted from given samples of rock using the Fischer assay, which is the standard tool for estimating oil yield. In the Fischer assay, a rock sample is heated at a constant rate to a target temperature—at 12° C per minute to 500° C, or 22° F per minute to 930° F— and then held at that temperature for 40 minutes. The test mimics the pyrolytic conditions present in some surface retorts, and in fact it is the benchmark for comparing the efficiency of various retort types. The assay does not measure the yield of gas from the pyrolytic process, which can be substantial.

Estimates of in-place oil resources do not take into account how much of the oil contained in the shale deposits can actually be recovered. Actual recovery rates depend upon the technology used to extract the oil on an industrial scale as well as features of each individual oil shale deposit and the local operational environment. For this reason, figures on total resources tend to be much larger than the amount that can be recovered.

A useful example would be the GRF. The GRF is estimated to contain more than 4 trillion barrels, or more than 600 million metric tons, of oil. However, considering current extraction technology, the location of the deposits, and many other economic and regulatory cost factors, it is unlikely that rock yielding less than about 15 gallons per short ton (63 litres per metric ton) could be processed economically from the GRF. This grade cutoff of 15 gallons per short ton effectively reduces the oil resources of the GRF by almost 75 percent, to between 1 and 1.5 trillion barrels (about 140 and 200 million metric tons). If technological and market conditions forced the grade cutoff upward to 25 gallons per short ton (104 litres per metric ton), then the actual recoverable amount would be reduced by more than 90 percent, to less than 500 billion barrels (60 million metric tons).

Because data for making such practical calculations are not available for other major oil shale basins of the world, resources are indicated on the basis of estimated oil content using the Fischer assay.

Recovery of Oil from Oil Shale

Minimum Organic Requirement

As is stated above, the organic matter present in oil shale is principally kerogen; no oil and little extractable bitumen is naturally present in oil shale. The kerogen found in oil shale is not distinct from the kerogen of petroleum source rocks—that is, the material from which petroleum was generated under conditions of heat and pressure over long periods of geologic time. To some extent the pyrolysis process for extracting oil from oil shale is comparable to the natural processes that generated conventional crude oil.

In order to be of commercial interest, oil shale must contain a large amount of organic matter—significantly larger than the 2 percent or more of organic carbon commonly found in the source rock from which conventional oil or gas may be generated. At the very least, the organic matter in a prospective oil shale must provide more energy than is required to process the shale. For instance, under controlled laboratory conditions, if the kerogen content of the shale is less than about 3 percent by weight, then its total calorific value will be needed simply to heat the rock to surface retorting temperatures and react the kerogen to oil and gas. Commercial conditions are far less efficient: heat is lost at various parts of the process, and other energy inputs are required for handling, upgrading, and so on. Consequently, the organic content of commercial-grade oil shales must be considerably higher than 3 percent.

In commercial practice, the actual energy recoverable from the shale—and, hence, the specific break-even point in organic content—is strongly dependent upon properties of the oil shale as well as features of the process. For example, the organic matter content of prospective western and eastern U.S. oil shale is roughly the same (7.5–15 percent). However, the organic matter in western shale is "richer" in hydrogen than eastern shale, yielding 20 to 40 gallons of oil per short ton (84 to 168 litres per metric ton) compared with only 10 to 15 gallons per short ton (42 to 63 litres per metric ton) in the "leaner" eastern shale. At the same time, western oil is relatively high in paraffinic compounds, so that, with upgrading, it becomes an excellent refinery feedstock that is well suited to large yields of diesel and jet fuel. Eastern shale oil, on the other hand, contains more aromatic compounds and, when upgraded, is better suited as a feed for catalytic crackers in the production of gasoline. Here the different end-products will determine the ultimate economic value of producing oil from either shale, so that the two different deposits may well have different cutoff grades.

Economic conditions elsewhere in the world may make it feasible to recover oil from leaner deposits than those found in the United States. In all cases, the overall energy balance is a critical determinant of whether shale oil production can proceed. Much research focuses on better defining this balance as well as looking for ways to improve it.

Pyrolysis

The technology for producing oil from oil shale is based on pyrolysis of the rock. Applied heat breaks the various chemical bonds of the kerogen macromolecules, liberating small molecules of liquid and gaseous hydrocarbons as well as nitrogen, sulfur, and oxygen compounds. Pyrolysis can be done aboveground (ex situ) in retorts, which are specially designed vessels that allow rapid heating of the rock in an oxygen-free environment. Under such conditions the pyrolytic reactions occur at temperatures in the range of 480–550° C (900–1,020° F). Surface retort hydrocarbon products typically contain relatively high proportions of olefins and diolefins, as well as sulfur and nitrogen compounds.

Pyrolysis can also be done by heating the rock underground (in situ). Because rock is an excellent insulator, heating rock formations underground in order to maximize production is a slow process, involving months to years. Under conditions of slow heating, the pyrolytic reactions occur at lower temperatures, roughly 325–400° C (620–750° F), and produce a lighter oil and a higher gas-to-oil ratio.

A third approach involves the creation of large surface capsules of tailored earth materials containing mined oil shale. A pit is excavated, lined with some type of engineered material to prevent escape of the products, and then filled with oil shale. At intervals in the fill, heating and drainage pipes and sensors are laid out, and the filled capsule is capped with impermeable material and soil. Hot gases are circulated through the pipes, and the products are extracted mainly as a vapour. This hybrid approach produces oil and gas similar to the in situ processes but in a shorter time.

Many specific pyrolytic processes have been developed. Whether the technologies are applied aboveground or underground, all of them fall into a relatively small number of basic methods based on their heating approach. Each method has its advantages and disadvantages.

- Internal-combustion approaches burn either gases or a portion of the shale to generate the heat for pyrolysis. This heat is transferred to the ore by the hot gas. Internal-combustion technologies have been designed for use in aboveground retorts as well as in situ. Three technologies that use this approach are the Kiviter process, employed in Estonia; the Fushun process of China; and the Paraho Direct process, designed in the United States.

- Hot-recycled-solids methods circulate either burned shale or an inert material as the heat carrier. Spent shale, which has had oil and gas removed from it, still has energy available in the carbon-rich char that is left behind on the mineral ash. Some technology options can burn this residual carbon to provide the heat for the process, which increases the effective utilization of the resource. The various hot-recycled-solids processes are applied only aboveground; they include the Estonian Galoter and Enefit 280 processes and the Canadian Alberta Taciuk Process.

- Methods that use conduction through a wall provide heat electrically or by burning a fuel outside the retort wall. They are applied both aboveground and in situ. The old Pumpherston process, used in Scotland beginning in 1862, involved external heating through the wall of the retort. This process was widely employed with various refinements introduced later in continental Europe. Modern technologies employing conduction through a wall are the Combustion Resources and Ecoshale In-Capsule processes, both designed in the United States.

- Externally generated hot gas methods inject a remotely heated gas into the retort zone. This has been done both aboveground and in situ, though the most prominent technologies are the Brazilian Petrosix process and the American Paraho Indirect process, both employed in aboveground retorts.

- Reactive fluids work in much the same manner as externally heated gas, but with a chemically reactive fluid such as high-pressure hydrogen. Hydrogen also partly upgrades the oil by removing sulfur and stabilizing reactive hydrocarbons. Reactive fluid technologies have been designed for aboveground and in situ use.

- Volumetric heating methods operate in much the same way as a microwave oven, emitting electromagnetic radiation or electric current that excites molecules in the rock and generates heat. Volumetric heating processes have been designed only for in situ use.

Surface Processing

Four basic steps are involved in the aboveground processing of oil shale: mining the ore (either through underground or surface mining), crushing the ore to a size that can be handled by the retort, retorting (heating) the crushed shale to pyrolysis temperatures, and upgrading the oil obtained by pyrolysis of the organic content of the shale.

Surface retorts may be classified by whether the reacting shale moves vertically through a stationary retort or horizontally, generally through a rotating drum-type retort. In addition, retorts are classified by whether they process shale as lumps (pieces varying from about 15 to 70 mm [0.5 to 2.75 inches] in size) or as fines (particles less than 10 mm [0.4 inch] in size).

In situ Processing

In situ processing differs from aboveground processing in that retorting to produce oil and gas takes place underground. No in situ processes are in use on a commercial scale, but several companies have investigated methods for heating large volumes of oil shale in place and extracting the oil and gas products using more-or-less traditional oil and gas wells. The most important aspect of in situ processes is the means of heating the rock and of containing the reaction products. Two promising systems would use electric heat to pyrolyze the rock. One would use a large array of vertical or horizontal wells with electrical heaters in them. The other would begin by drilling parallel horizontal wells, hydraulically fracturing the rock, and then injecting an electrically conductive propping medium into the fracture system. A single horizontal well drilled at a right angle to the parallel wells would connect them, and electric current would be passed through this circuit, essentially creating a large platelike heating element underground to heat the rock. Other heating methods would use downhole burners or injection of surface-heated gases such as air or carbon dioxide.

Upgrading

The product of most surface retorts is relatively dense oil consisting of large hydrocarbon molecules. Also, shale oil is commonly high in compounds containing oxygen, sulfur, or nitrogen, impurities that can degrade refinery equipment or, if present in the end product, create noxious pollution upon combustion. For those reasons, oil derived from shales must be upgraded if its use is to be extended beyond heating oil and bunker fuel. For example, removal of particulates and diolefins from GRF shale oil reduces the fouling of equipment, and treatment of the high paraffin content through hydrotreating yields high-quality oil that can be refined into products such as jet fuel and diesel fuel. Another upgrade likely to be required upstream of the refinery is the removal of nitrogen, which is known to foul refining catalysts. All upgrading processes require significant effort and expense, but they are well within the realm of existing technology.

Oil produced by in situ processes is generally much lighter and is likely to require less upgrading than oil produced in aboveground retorting. Nevertheless, various upgrading steps are required, including hydrotreating, in part using the higher hydrogen content of the wellhead product. In addition, gas produced from in situ retorting is likely to be "sour"—that is, containing hydrogen sulfide and carbon dioxide, and those impurities too must be removed.

Environmental Issues

The production of oil from shales has a potentially serious impact on the environment. Four specific areas of concern dominate discussion regarding development of the resource: greenhouse gas output, water consumption and pollution, surface disturbance, and socioeconomic effects.

Because oil and gas are produced by heating oil shale and because heating methods typically involve hydrocarbon combustion either at the site or in power plants nearby, shale processing inevitably results in the emission of carbon dioxide (CO_2), the most common greenhouse gas. It is commonly estimated that in situ processes, if applied on a commercial scale, would emit at least 10 to 20 percent more CO_2 than conventional petroleum production. Some aboveground processes, which operate at higher temperatures and break down carbonate minerals, may produce 50 percent more CO_2 than conventional oil processes. A number of options have been proposed to reduce those emissions. For instance, it has been suggested that CO_2 could be captured and sequestered in previously produced in situ blocks, or it could be piped to conventional oil fields for use in enhanced oil production.

The pyrolytic production of oil from rock does not consume water. However, full-scale oil shale processing is expected to require 0.7 to 1.2 litres of water for every litre of oil produced, primarily for site remediation, drilling or mining, and upgrading of the synthetic crude. Power generation could be an important additional application for oil shale if traditional methods of steam condensation are used, but most developers plan to use air cooling in water-constrained areas. Although water use associated with full-scale oil shale processing appears reasonable when compared with the much greater usage of water by various biofuel processes, attempts are being made to reduce water use. A very large oil shale industry, producing 500,000 barrels per day of shale oil, would still account for much less than 1 percent of Colorado's total water usage in a year.

The contamination of surface water or groundwater by mining and retorting operations can be prevented or mitigated by applying established best industrial practices. In situ retorting leaves behind some products and by-products, including organic compounds that could contaminate groundwater reservoirs. In addition, alteration of the underground rock by heating may make certain inorganic contaminants more mobile in groundwater. Experiments have been conducted showing that those constituents can be removed effectively prior to abandonment of a site. Nevertheless, potential water contamination by in situ operations is an important concern. Finally, the management of spent shale piles and the reclamation of mined and developed areas would require the use of water—both a technical challenge and a sociopolitical issue in arid regions.

In huge open-pit or underground operations, large amounts of rock material have to be moved in order to provide shale for surface retorting. Such operations can adversely affect the integrity of the land, grazing and agricultural activities, and local fauna and flora. Even in situ production using boreholes may have a significant surface impact, though the concentration of some resources in relatively small areas may help to keep the overall footprint relatively modest. Mine reclamation and restoration is feasible and has been demonstrated in past efforts at oil shale exploitation.

Development of oil shale on an industrial scale would inevitably affect the rural regions where the resources occur. For instance, before royalty revenues arrive from the new development, local governments might experience difficulty in meeting the demand for increased services and infrastructure. Oil shale developers would have to address such concerns, as well as the anxiety of people who have experienced the rise and fall of previous oil shale ventures, in order to acquire a mandate to continue their efforts.

Petroleum Seeps

Petroleum seeps are naturally occurring springs where oil leaks out of the ground. These seeps occur when cracks in the earth allow the oil to escape, and pool above the surface.

Places where Petroleum Seeps are Found

Petroleum seeps are typically found where oil drilling and excavation activities are being carried out. This is thought to be due to the nature of excavation which disturbs the surrounding areas, and can cause fissures in the ground.

Seeps can also occur where the land has moved over time. In this scenario, the folds of sedimentary rock shift, which allows the oil to leak out to form a natural seep on the surface.

These seeps tend to occur in clusters around the world, and many can be found in the Gulf of Mexico and off the southern coast of California.

Composition of Petroleum Seeps

The oil that makes it to the surface to form a seep will eventually convert from a clear-like fluid into a tar-like substance called asphaltum. This is because the lighter elements of the oil evaporate, leaving the heavier oil compounds which then oxidise.

This oxidised substance is also attacked by bacteria and in its final stage becomes sticky and black.

Toxicity of Petroleum Seeps

Whenever there is a news report on an oil spill, the concern is always focussed towards the environment and damaging effects on wildlife. So shouldn't we be as worried about petroleum seeps?

The answer is no. This is because seeps are typically very old, and take a long time to emerge at the surface. The oil that finally emerges is not pure crude oil, as it has been already heavily biodegraded by bacteria, deep beneath the surface of the earth.

The substance that does come out is still toxic. However, because seeps have been around for a very long time, organisms and wildlife have managed to build up their defences. Most who live in the area of seeps have adapted well to these toxic compounds. Even more surprising is that some unique species of wildlife actually use the chemicals and hydrocarbons released in a seep for metabolic energy.

Things to be Learned from Petroleum Seeps

In contrast to petroleum seeps, the extraction and transportation of crude oil typically produces high-volume quantities in a short period of time. It also takes place in areas that have not had a long exposure to these types of chemical compounds. This means that seeps are generally looked upon by scientists as a rare chance to study how natural processes affect oil, and how species adapt over time to toxic chemicals.

It is hoped that these studies may provide better techniques in the future for dealing with actual oil spills, and a greater understanding on how they affect wildlife and the environment.

Basin and Petroleum System Modeling

Basin and petroleum system modeling allows geoscientists to examine the dynamics of sedimentary basins and their associated fluids to determine if past conditions were suitable for hydrocarbons to fill potential reservoirs and be preserved there.

The best way to reduce investment risk in oil and gas exploration is to ascertain the presence, types and volumes of hydrocarbons in a prospective structure before drilling. Seismic interpretation can delineate closed structures and identify potential subsurface traps, but it does not reliably predict trap content. Drilling on a closed structure, even near a producing oil or gas field, holds no guarantee that similar fluids will be found. Profitable exploration requires a methodology to predict the likelihood of success given the available data and associated uncertainties.

The concept connects the past—a basin, the sediments and fluids that fill it, and the dynamic processes acting on them—to the present: hydrocarbon discoveries. Early endeavors sought to describe how basins form, fill and deform, focusing mainly on compacting sediments and the resulting rock structures. Subsequent efforts concentrated on developing methods to model these processes quantitatively. This area of study, which has become known as basin modeling, applies mathematical algorithms to seismic, stratigraphic, paleontologic, petrophysical, well log and other geologic data to reconstruct the evolution of sedimentary basins.

Geochemists developed methods for the prediction of the petroleum generation potential of a lithologic unit in quantitative terms. Soon after, they began to use sedimentary basin models as structural frameworks for geochemical genetic correlations between hydrocarbons and source rocks. A number of scientists worked on the notion independently, so several names were given to the idea, including oil system, hydrocarbon machine, petroleum system and independent petroliferous system; each approach emphasized different aspects of this multifaceted problem. The term "petroleum system" is now commonly used within the industry, and the concept it describes synthesizes many features of the collective work.

A petroleum system comprises a pod of active source rock and the oil and gas derived from it as established by geochemical correlation. The concept embodies all of the geologic elements and processes needed for oil and gas to accumulate. The essential elements are an effective source rock, reservoir, seal and overburden rock; the last facilitates the burial of the others.

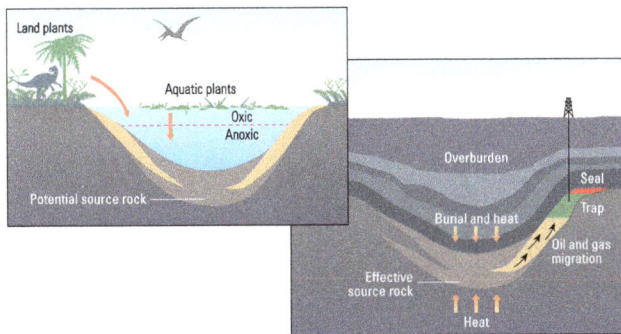

Simulating geologic, thermal and fluid-flow processes in sedimentary basins over time. Basin and petroleum system modeling (BPSM) reconstructs the deposition of source, reservoir, seal and overburden rocks and the processes of trap formation and hydrocarbon generation, migration and accumulation from past (left) to present (right).

The processes include trap formation and the generation, migration and accumulation of petroleum. These elements and processes must occur in the proper order for the organic matter in a source rock to be converted into petroleum and then to be stored and preserved. If a single element or process is missing or occurs out of the required sequence, a prospect loses viability.

Basin and petroleum system modeling (BPSM) tracks the evolution of a basin through time as it fills with fluids and sediments that may eventually generate or contain hydrocarbons (left). In concept, BPSM is analogous to a reservoir simulation, but with important differences. Reservoir simulators model fluid flow during petroleum drainage to predict production and provide information for its optimization. The distance scale is meters to kilometers, and the time scale is months to years. Although the flow is dynamic, the model geometry is static, remaining unchanged during the simulation. On the other hand, BPSM simulates the hydrocarbon-generation process to calculate the charge, or the volume of hydrocarbons available for entrapment, as well as the fluid flow, to predict the volumes and locations of accumulations and their properties. The distance scale typically is tens to hundreds of kilometers, and the periods covered may reach hundreds of millions of years. The model geometry is dynamic and often changes significantly during simulation.

Basin and petroleum system modeling brings together several dynamic processes, including sediment deposition, faulting, burial, kerogen maturation kinetics and multiphase fluid flow. These processes may be examined at several levels, and complexity typically increases with spatial dimensionality; the simplest, 1D modeling, examines burial history at a point location. Two-dimensional modeling, either in map or cross section, can be used to reconstruct oil and gas generation, migration and accumulation along a cross section. Three-dimensional modeling reconstructs petroleum systems at reservoir and basin scales and has the ability to display the output in 1D, 2D or 3D, and through time. Most of the following discussion and examples pertain to 3D modeling; if the time dimension is included, the modeling can be considered 4D.

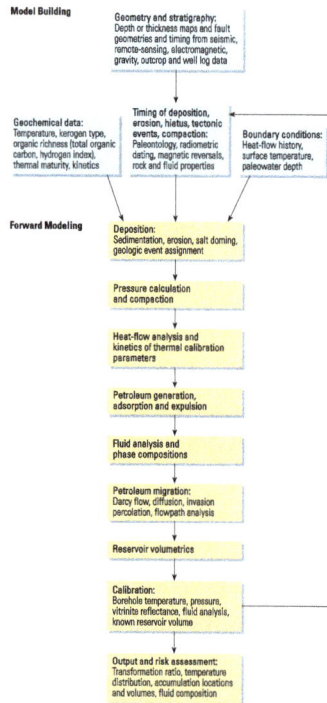

The multiple and interrelated steps of BPSM.

Basin and petroleum system modeling consists of two main stages: model building and forward modeling. Model building involves constructing a structural model and identifying the chronology of deposition and physical properties of each layer. Forward modeling performs calculations on the model to simulate sediment burial, pressure and temperature changes, kerogen maturation and hydrocarbon expulsion, migration and accumulation. Calibration compares model results with independent measurements to allow refinement of the model.

For any spatial dimensionality, BPSM performs deterministic computations to simulate the history of a sedimentary basin and its associated fluids. The computations require a model, or a discretized numerical representation of layers containing sediments, organic matter and fluids with assigned properties. A model is constructed from geophysical, geologic and geochemical data. The layers are subdivided into grid cells within which properties are uniform. Computer programs simulate physical processes that act on each cell, starting with initial conditions and progressing by a selected time increment to the present. Model outputs, such as porosity, temperature, pressure, vitrinite reflectance, accumulation volume or fluid composition, can be compared with independent calibration information, and the model can be adjusted to improve the match.

Basin and petroleum system modeling is an iterative process with many interrelated steps, each of which is a scientific discipline in itself (right). Assembling these steps in a single workflow is a daunting task. A few major oil companies and a handful of contractor companies have developed systems that perform these calculations in one way or another. The Schlumberger approach combines Petrel seismic-to-simulation software for building the basin model with PetroMod petroleum system modeling software for simulating the generation, migration and accumulation of hydrocarbons. The following explanation of BPSM describes aspects of the general process along with a few features particular to PetroMod software.

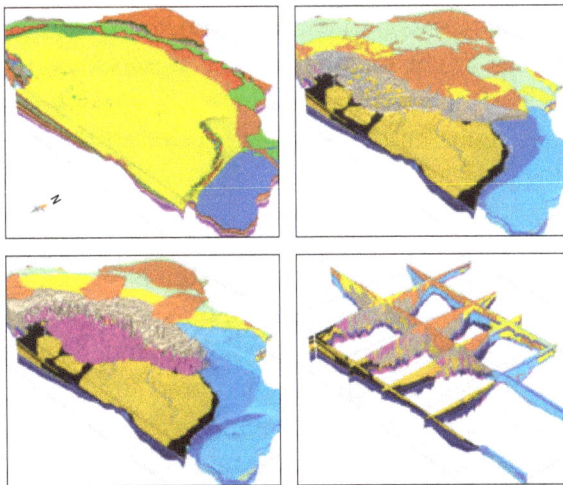

A regional-scale structural model of the entire northern Gulf of Mexico.

In general, a preparation step before modeling defines the petroleum system to be modeled. Formally, the name of a petroleum system consists of the name of the active source rock, followed by a hyphen and the name of the reservoir rock that contains the largest volumes of petroleum from the source rock. The name ends with a punctuation symbol in parentheses that expresses the level of certainty—known, hypothetical or speculative—that a particular pod of active source rock has generated the hydrocarbons in an accumulation. In a known petroleum system, the active source rock has a clear geochemical match with trapped hydrocarbons. For example, in the Shublik-Ivishak (!) petroleum system of the North Slope of Alaska, geochemical analysis has determined the Triassic Shublik Formation source rock is the source of hydrocarbons in the Triassic Ivishak reservoir. The (!) indicates it is a known petroleum system. In a hypothetical petroleum system, designated by the (.) symbol, the source rock has been characterized by geochemical

BP combined local and regional maps of salt and sediment horizons to construct a regional model covering approximately 1.1 million km^2 (400,000 mi^2) that accounts for complex salt movement. Each colored layer (top left) represents a stratigraphic interval of specified age. Colors in the top right correspond to different sedimentary depositional settings and mixtures of rock types. The bottom left image displays the model with shallow horizons removed to expose the allochthonous salt (magenta). The fence diagram (bottom right) shows the internal detail of the model, including multiple salt layers. All images represent the present-day geology analysis, but no match has yet been made with a hydrocarbon accumulation. In a speculative petroleum system, labeled with (?), the correlation of a source rock to petroleum is merely postulated based on geologic inference.

The first step is to create a depth-based structural model of the area of interest, which may encompass a single petroleum system in a small basin or multiple petroleum systems in one basin or many basins across a region. Input is typically in the form of formation tops and layer thicknesses and can be imported from a separate model-building program. Data sources might include seismic surveys, well logs, outcrop studies, remote-sensing data, electromagnetic soundings and gravity surveys. This model of present-day architecture represents the final result of all the processes acting on the basin throughout geologic time.

The modeler must then analyze to the present day geometric model to describe the deposition chronology and physical properties of the basin fill materials and to identify postdepositional processes—an undertaking that will enable reconstruction of the basin and its layers and fluids throughout geologic time. This analysis establishes a basin history that is subdivided into an uninterrupted series of stratigraphic events of specified age and duration. These events are summarized in a petroleum system events chart. Each event represents a span of time during which deposition, nondeposition or erosion occurred. This summary describes the chronology of the geologic elements in a petroleum system. Syn- and postdepositional episodes of folding, faulting, salt tectonics, igneous intrusion, diagenetic alteration and hydrothermal activity can be included to explain the model. Determining the timing of trap formation and of the remaining processes—generation, migration and accumulation of hydrocarbons—is one of the main goals of BPSM.

An important concept in process timing is the "critical moment." This is the time of generation, migration and accumulation of most of the hydrocarbons in a petroleum system. The critical moment occurs in the range of 50 to 90% transformation ratio (TR), which is the relative conversion of source-rock organic matter to hydrocarbons. The selection of the time within this range is at the discretion of the modeler.

The absolute age of each layer in the basin and petroleum system model is an important parameter for determining the timing of the processes that generate, move and trap petroleum. Age information may be available from paleontologic data, radiometric dating, fission-track dates and magnetic-reversal tracking. In many basins, known petroleum source rocks have been assigned to global geologic periods based on geochemical and biostratigraphic determinations.

Identification of the lithology and depositional environment of each stratigraphic unit is crucial. For example, classifying the depositional environment, and thus properties, such as porosity and permeability, of coarse-grained sediments helps identify their potential as reservoir or carrier rocks that facilitate migration of petroleum from source rock to reservoir. Characterizing the source-rock depositional environment helps predict the probable petroleum product generated through kerogen maturation. Fine-grained sediments deposited in deep marine basins, on continental shelves and in anoxic lakes all contain different types of kerogen, leading to different petroleum outputs.

Source-rock properties are needed as inputs to simulate the reactions that govern the degradation of organic material to produce hydrocarbons. These essential properties are the total organic carbon (TOC) measured by combustion of rock samples and the hydrogen index (HI) obtained through pyrolysis of rock samples for petroleumgeneration potential.18 Also required are kinetic parameters for the thermal conversion of the source-rock kerogen to petroleum. Another measure of kerogen maturity is vitrinite reflectance. As an independent measurement that is not a PetroMod input, it provides a means to calibrate the model output. Simulation of the burial history can be used to predict the expected vitrinite reflectance at any depth or time in the model. Calibration entails adjusting the model so that the simulated vitrinite reflectance matches that measured in samples at varying depths in the well.

Several other physical properties must also be specified for each layer. Porosity and permeability in reservoir and carrier layers are important for fluid-flow computations and reservoir volumetric estimates. Permeability of source rocks affects the efficiency with which generated hydrocarbons can be expelled. Heat capacity and thermal conductivity, usually inferred from lithology and mineralogy, are needed for the thermal calculations that model kerogen maturation and petroleum generation. In addition, density and compressibility data are required inputs to model compaction and burial.

The burial history of basin sediments contains information about burial depth and preservation of organic material, which are related to the temperatures and pressures the sediments were exposed to and the durations of exposure. Temperature is the primary variable in conversion of kerogen to petroleum, and pressure is important for migration of fluids. Key inputs for building a burial history include sedimentation rate, compaction, uplift, erosion and depositional environment.

The thermal history of a basin is linked to the history of the crust in which it formed. Crustal behavior determines basin subsidence, uplift and heat flow. Modeling the petroleum potential of a basin requires reconstruction of the temperature over geologic time and across the basin. Therefore, in addition to model properties, some specific past conditions must be evaluated. These conditions, treated as boundary conditions by the modeling software, include paleobathymetry, which determines the location and type of deposition. Other boundary conditions are sediment/water interface temperatures throughout geologic time which, along with paleoheat-flow estimates, are required to calculate the temperature history of the basin.

Fast Forward

After the boundary conditions and ages and properties of all layers have been defined, the simulation can be run forward, starting with sedimentation of the oldest layer and progressing to the present. The following steps summarize the workflow of the Petro-Mod modeling software.

Deposition—Layers are created on the upper surface during sedimentation or are removed during erosion. Depositional thickness, which may have been greater than current thickness, can be calculated by several methods: porosity-controlled backstripping starting with present-day thickness, importation from structural-restoration programs, and estimation from sedimentation rate and depositional environment.

Generic Events Chart

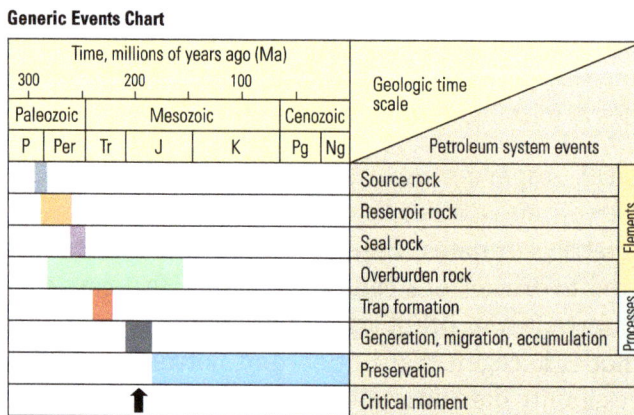

An events chart depicting timing of a petroleum system.

Each of the colored horizontal bars represents the time span of an event. For this system, all the essential elements and processes are present and the timing is favorable; source-rock deposition was followed by deposition of reservoir, seal and overburden rock. Also, the trap formed before hydrocarbon generation, migration and accumulation. Because the reservoir was filled before the end of Jurassic time, its hydrocarbons must be preserved for more than 180 million years to remain a viable prospect. The critical moment (black arrow) was chosen to be approximately half-way through the period of hydrocarbon generation, migration and accumulation.

Pressure calculation and compaction—The pressure calculation treats dewatering as a onephase flow problem driven by changes in overburden weight caused by sedimentation. In addition, internal pressure-building processes such as gas generation, quartz cementation and mineral conversions can be taken into account. Compaction causes changes in many rock properties, including porosity, and to a lesser extent, density, elastic moduli, conductivity and heat capacity. Therefore, pressure and compaction calculations must be performed before heat-flow analysis in each time step.

Heat-flow analysis—The goal of heat-flow analysis is temperature calculation, a prerequisite for determining geochemical reaction rates. Heat conduction and convection from below as well as heat generation by emissions from naturally occurring radioactive minerals must be considered. Incorporating the effects of igneous intrusions requires the inclusion of thermal phase transitions in sediments. Thermal boundary conditions with inflow of heat at the base of sediments must also be formulated. These basal heat-flow values are often predicted using crustal models in separate preprocessing programs or are interactively calculated from crustal models for each geologic event.

An example of the level of complexity involved in heat-flow analysis is seen in a study of petroleum systems in the San Joaquin basin in California, USA. The process begins with the present-day latitude of the basin. A PetroMod option recreates the plate tectonic locations of the basin through time and calculates the corresponding temperatures of the sediment/water interface. These surface temperatures are then corrected for water depth to give past sediment/water interface temperatures. These constrain the paleoheat-flow profiles.

Present-day heat-flow values were estimated using temperature and thermal-conductivity data from wells and aqueduct tunnels in the San Joaquin basin. The temperatures—measured as functions of depth—were used to determine the geothermal gradient. Contemporary heat flow was calculated by multiplying the geothermal gradient by the thermal conductivity. The resulting map of surface heat flow was an input to the PetroMod software, which delivered sourcerock maturity values that matched available maturity measurements.

Petroleum generation—The generation of petroleum from kerogen in source rocks, called primary cracking, and the subsequent breakdown of oil to gas in source or reservoir rocks, called secondary cracking, can be described by the decomposition kinetics of sets of parallel reactions. The number of chemical components produced in most models can vary between 2 (oil and gas) and 20. The cracking schemes may be quite complex when many components and secondary cracking are taken into account. PetroMod software uses a database of reaction kinetics to predict the phases and properties of hydrocarbons generated from source rocks of various types. In addition, adsorption models describe the release of generated hydrocarbons into the free pore space of the source rock.

Fluid analysis—The generated hydrocarbons are mixtures of chemical components. Fluid-flow models deal with fluid phases that are typically liquid, vapor and supercritical or undersaturated phases. The fluid-analysis step examines temperature- and pressure-dependent dissolution of hydrocarbon components in the fluid phases to determine fluid properties, such as density and viscosity, for input to fluid-flow calculations. These properties are also essential for subsequent migration modeling and calculation of reservoir volumetrics. Fluids may be modeled using a black-oil model, which has two components or phases, or a multicomponent model.

Fluid-flow calculations—There are several fluid-flow approaches to model migration of generated hydrocarbons from source rock to trap. Darcy flow describes multicomponent three-phase flow based on the relative-permeability and capillary pressure concept. With this method, migration velocities and accumulation saturations are calculated in one step. Describing fluid migration across faults requires special algorithms.

A simplified fluid-flow calculation is made by flowpath analysis. In high-permeability layers, known as carriers, lateral petroleum flow occurs instantaneously on geologic time scales. It can be modeled with geometrically constructed flowpaths to predict the locations and compositions of accumulations. Spilling between and merging of drainage areas must be taken into account. In a hybrid method, flowpath analysis in high permeability zones may be combined with Darcy flow in low-permeability regions.

Alternatively, migration and accumulation can be modeled by invasion percolation in the PetroMod software. This calculation assumes that on geologic time scales petroleum moves instantaneously through the basin driven by buoyancy and capillary pressure. Any timing constraint is neglected, and the petroleum volume is subdivided into small finite amounts. Invasion percolation is convenient for modeling fluid flow in faults. The method is especially efficient for one-phase flow consisting of only a few hydrocarbon components and for the introduction of higher-resolution migration.

Reservoir volumetrics—The height of a petroleum accumulation is limited by the capillary entry pressure of the overlying seal and the spill point at the base of the structure. Loss at the spill point and leakage through the seal reduce the trapped volume. Other processes, such as secondary cracking or biodegradation, can also impact the quality and quantity of accumulated petroleum.

Calibration parameters—It is possible to predict rock temperature, vitrinite reflectance values and concentration ratios of molecular fossils (biomarkers) using models based on Arrhenius-type reaction rates and simple conversion equations. These temperature-sensitive predictions can be compared with measured data to calibrate uncertain thermal input data, such as paleoheat-flow values.

Risk—Numerical models, including basin and petroleum system models, provide scenarios for what might happen given various constraints on the input data. The impact of uncertain data can be studied by multiple simulation runs with varying model parameters. Assignment of varying parameters and the corresponding impact on the model can be performed with statistical methods, such as Monte Carlo simulations. These simulations do not provide a unique answer, but rather a range of possible outcomes with estimates of uncertainty. Increased computing power combined with multiple simulations allows the user to compare the effects of various scenarios and identify which variables exert the most control on the computed results. Final outcomes are mainly scenario probabilities and confidence intervals—for example, percentiles limiting in situ petroleum volumes.

Global Mean Surface Temperature

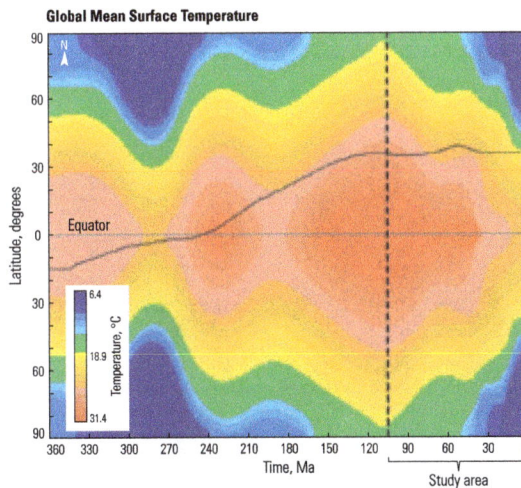

Estimated global mean surface temperature throughout geologic time.

Variations of this chart can be used to calculate the paleotemperature of the sediment/water interface for sediments deposited at any latitude and any age. The solid black line shows the variation in latitude with time for the San Joaquin basin study area. The dashed line represents the beginning of deposition of the sediments studied. The portion of the solid black line to the right of the dashed line tracks the temperature of the sediment/water interface during the study period. A PetroMod calculation corrects these temperatures for subsequent water depth.

Because of the highly sensitive nature of petroleum system modeling results, many companies keep BPSM success stories to themselves. An example from Indonesia was released because the operator sought drilling partners after a study showed that deepwater acreage on the Mahakam Delta and Makassar Slope off Kalimantan would likely produce oil, contrary to belief at the time that source rocks were gasprone and thermally post mature. The generally accepted geochemical-stratigraphic model for the area restricted the effective and oil-prone coaly source rocks to updip shelfal areas. Age equivalent rocks on the outer shelf were also thought to be buried too deeply to preserve good reservoir quality.

Before the deadline to relinquish the blocks, Mobil conducted a study of 61 oil samples provided by the principal operators in the area. Using biomarkers in the oil samples, re-interpreted sequence stratigraphy and proprietary kinetic parameters, Mobil geologists performed BPSM, which predicted that most of the Miocene source rock in the area of interest would be within the present-day oil window and would be currently generating hydrocarbons. Use of the model resulted in major oil discoveries by Mobil and its partner Unocal in the deepwater Makassar Straits, with some wells producing 10,000 bbl/d [1,600 m³ /d] of oil from areas previously considered nonprospective. The study also changed the way the industry views deepwater deltaic petroleum systems worldwide.

Historically, BPSM has been applied to basin-scale studies to assess uncertainties in hydrocarbon charge, migration and trap formation. Increasingly, it is being applied to understand the origins of fluid complexities in producing fields. The next two examples demonstrate how PetroMod simulations help explain fluid distributions that pose challenges to production.

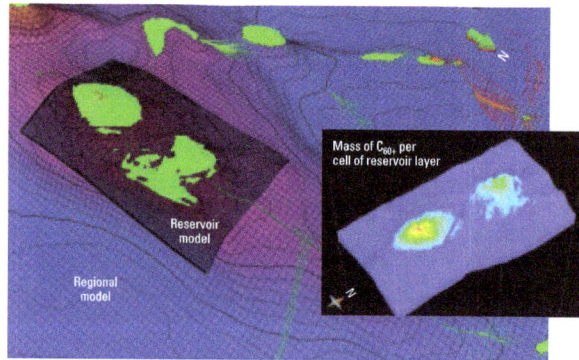

Linking models of different scales.

A portion of the 3D regional PetroMod model of a petroleum system in Kuwait was constructed on a coarse grid of 1,200 by 1,200 m [3,900 by 3,900 ft]. Depth to the reservoir is color-coded, with depth increasing from red to blue. Contour interval is 50 m [164 ft]. The 100- by 100-m [330- by 330-ft] grid cells from a Petrel reservoir model were included in the PetroMod model by means of local grid refinement. Green indicates oil accumulations and red indicates gas. Thin green and red lines show the multiple migration paths taken by the fluids to the traps. The inset depicts results of modeling the present-day distribution of the reservoir's dissolved heavy-oil (C60+) components. The color scale (not shown) is in megatons and ranges from 0 (blue) through yellow to 0.04 (red). The current distribution of heavy-oil components in each reservoir can be fully explained as a function of the generation, expulsion and migration history.

Petroleum System Modeling to Understand Production

Three-dimensional fluid-flow modeling can provide a competitive advantage at different times in the life of a field. Basin-scale petroleum system modeling is designed for use during exploration, and field-scale reservoir modeling is performed during production. Until recently, however, the vastly different scales of petroleum system and reservoir models have impeded progress toward linking these powerful methods. Working with the Kuwait Oil Company (KOC), Schlumberger used local grid refinement (LGR) to combine basin- and reservoir-scale models. While LGR is well-established in reservoir simulators, this was its first use in 3D fluid-migration simulation. Applied to an oil field in Kuwait, the technology improved understanding of the origin and distribution of heavy oil within the field and helped to assess the impact of these heavy-oil deposits on development strategies.

A PetroMod model on a regional scale helped quantify the location and timing of petroleum expulsion from source rock, volumes and composition of the products, and migration pathways. This exercise revealed that two effective post-salt source rocks, the Cretaceous Makul Formation and the Cretaceous Kazhdumi Formation, produced fluids that migrated along different pathways to the trap at different times, resulting in a complex filling history.

The PetroMod model incorporated high resolution grids from a Petrel reservoir model by means of the new local grid refinement option in PetroMod software. The linked models helped engineers investigate the influence of pressure-volume-temperature changes on discontinuous deposits of heavy oil throughout basin history—including multiple charging episodes and periods of uplift and erosion that led to tilting of traps and ancient oil/water contacts. The results from linking the models provided KOC with testable hypotheses on the mechanisms of heavy-oil formation, which should prove useful for predicting the distribution of these low permeability barriers for input to ECLIPSE reservoir simulation of field production.

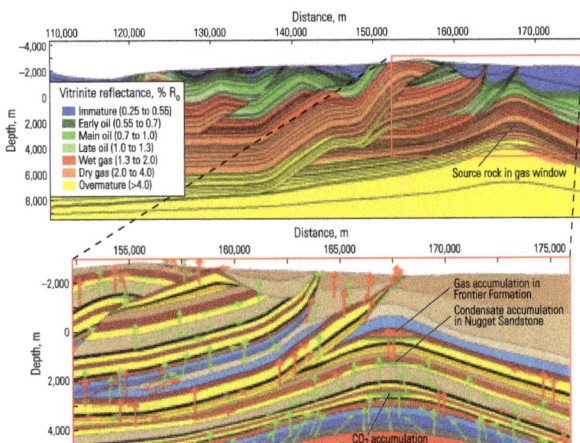

Modeling maturity and migration.

PetroMod software modeled the products of maturation from multiple source rocks in a complex thrust zone (top). Migration calculations over a portion of the section (bottom) predicted accumulation of CO_2 in a deep Paleozoic reservoir, condensate in the Nugget Sandstone and gas in the Frontier Formation. Green and red arrows represent flowpaths taken by liquid and vapor phases, respectively. Results matched published fluid data from the field.

Modeling Gas Migration in a Thrust Belt

Tectonically complex, compressional environments pose challenges to BPSM. Analysis of a petroleum system in such an area—a thrust belt in western Wyoming, USA—used public data and structural-restoration software to determine the distribution of gas, gas condensate and CO_2 in wells in the La Barge field.

Two overlapping tectonic phases, the Late Jurassic to Early Tertiary Sevier orogeny, and the Late Cretaceous to Early Neogene Laramide orogeny, contributed to structural complexity in the geology seen today. Geologists reconstructed the 90 million-year history of the basin using third-party software and input the resulting model into Petro-Mod software. The modeled present-day source-rock maturity was calibrated using temperature data from wells in the distant Wind River basin.

Petroleum migration simulated using a combination of Darcy flow and flowpath modeling tracked the movement of fluids to present-day accumulations. Predicted petroleum properties, such as API gravity and gas/oil ratio (GOR), match published data on fluids produced from the La Barge field.

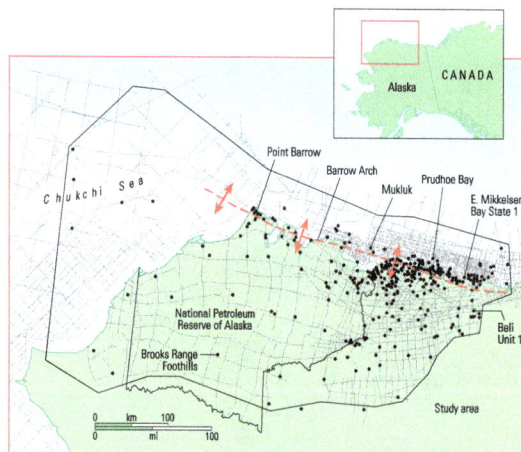

North Slope regional BPSM study.

The study area covered most of the National Petroleum Reserve of Alaska and the central North Slope, and extended over the eastern part of the Chukchi Sea. The red dashed line indicates the trace of the Barrow Arch. The red arrows point in the direction of the plunge.

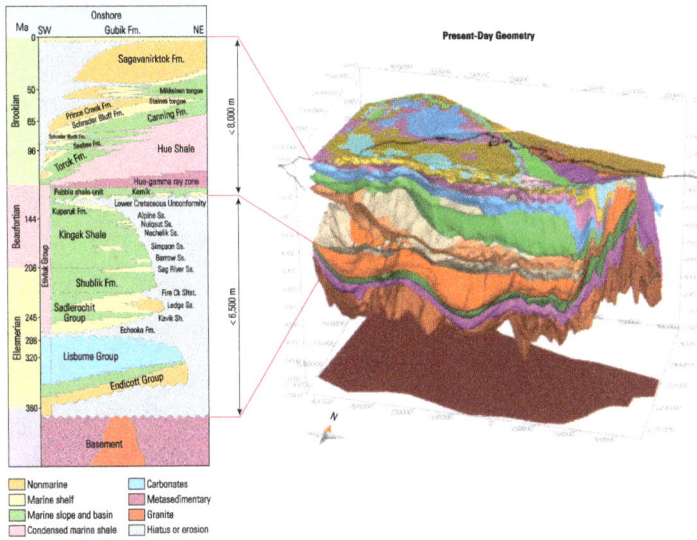

Alaska structural model.

The stratigraphy (left) catalogs the source rocks (Kekiktuk coals of the Endicott Group, Shublik Formation, Kingak Shale, Hue Shale, Hue-GRZ) and reservoir rocks (Kemik, Kuparuk Formation, Sag River Sandstone, Ledge Sandstone) of the North Slope petroleum systems. Stratigraphic reconstruction accounted for more than 4,000 m [13,000 ft] of eroded overburden, which significantly affected burial and maturation. The structural model (right) contains 44 layers and was constructed from well and seismic data. The black line near the top of the model is the present-day coastline. The Lower Cretaceous Unconformity (LCU) is a major structural feature that strongly influenced the migration and accumulation of petroleum.

North Alaska Petroleum Systems

In addition to improving the understanding of complex fluid distributions, basin and petroleum system modeling can be applied both to frontier provinces and to well-understood areas. An example from the North Slope unites these approaches by combining BPSM on a regional scale with prospect-scale modeling to help geoscientists understand the petroleum systems in a region spanning vast underexplored areas and those containing significant known reserves.

The study, undertaken by Schlumberger and the US Geological Survey (USGS), had multiple objectives: to use public geophysical, geologic and log data to develop a comprehensive model of depositional units and rock properties; to better define the timing of basin filling, source-rock maturation and petroleum migration and accumulation; and to quantify the volumes, compositions and phases of generated hydrocarbons.

The study area covers 275,000 km² [106,000 mi²] and includes data from more than 400 wells. The model, built from logs and 2D seismic data, featured a grid with 1- by

1-km [0.6- by 0.6-mi] resolution. In the western portion of the study area, in the Chukchi Sea, data are relatively sparse, whereas the eastern part of the study area is well-explored and contains several productive fields, including Prudhoe Bay, the largest field in North America.

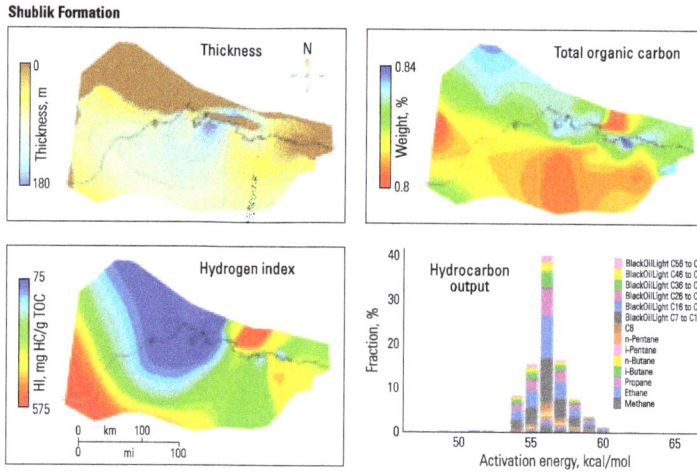

Shublik Formation source-rock properties.

Model input included source-rock thickness (top left), total organic carbon (top right), hydrogen index (bottom left) and expected hydrocarbon output based on reaction kinetics measured from an immature equivalent of the source rock in the Phoenix 1 well (bottom right). Similar input data were created for each of the oil-generating source rocks.

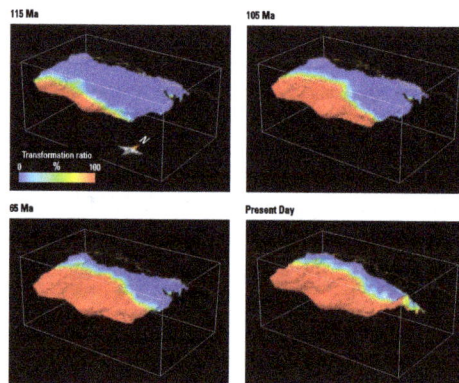

Snapshots of the source-rock transformation ratio of the Shublik Formation.

Kerogen in this source rock undergoes increasing transformation to petroleum as the layer is buried. The transformation ratio is color-coded from blue (0%) to red (100%). By 65 Ma, more than half the mapped Shublik Formation kerogen had undergone 100% transformation. The updip portion to the north along the Barrow Arch remained immature. The present-day plot shows that the formation is currently undergoing transformation as it is buried in the northeast.

The present-day subsurface geometry shows complex stratigraphy. Hydrocarbons from five source rocks have accumulated in several reservoir formations, creating multiple petroleum systems. Important source rocks lie beneath a prominent stratigraphic boundary: the Lower Cretaceous Unconformity (LCU). Tracking the deposition of the overburden rocks, called the Brookian foresets, facilitates an understanding of the burial history and helps determine the maturation of source rock. Snapshots of PetroMod models through geologic time show the burial of the Triassic, Jurassic and Cretaceous source rocks and the progradation of the Brookian foresets from southwest to northeast and their eventual erosion. This dynamic model formed the structural input to the PetroMod model.

For each of the source rocks, input included maps of layer thickness, original TOC and original HI, and expected hydrocarbon output based on kinetic measurements from thermally immature source-rock samples at different well locations.

Among the results of PetroMod modeling are time-lapse maps of the transformation ratio, or percentage of kerogen transformed into petroleum, for each source rock. In general, as burial depth increases, more of the source rock passes through the oil-generation window, allowing more-complete maturation of the organic matter. Most of the kerogen in the Shublik Formation outside the Barrow Arch has already undergone 100% transformation to petroleum.

The results of BPSM can be calibrated by comparison with independent information on basin history and kerogen maturation. Two key calibration parameters are temperature and vitrinite reflectance measured in boreholes and from borehole samples, respectively.

Incorporating burial pressure, heat-flow calculations, kinetics of thermal maturation and multiphase flow simulations, PetroMod software modeled the expulsion of liquid and vapor hydrocarbons from the many source rocks and the migration of these fluids to trapping structures. Tracking fluid migration to the present indicates areas where hydrocarbons have accumulated.

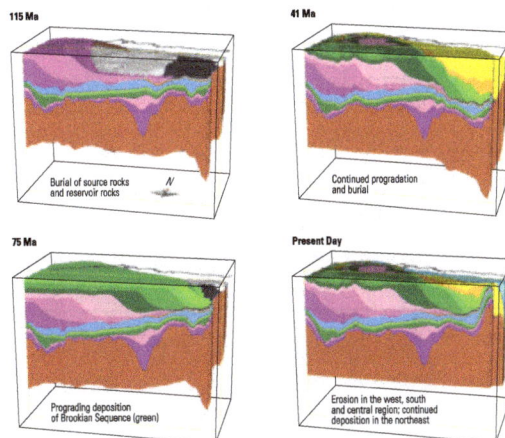

Time-lapse basin models.

Forward modeling begins prior to 115 million years ago (Ma) and proceeds to the present day. The snapshots show southwest to northeast progradation of the Brookian Sequence that significantly influenced the timing of petroleum generation and migration from the underlying source rocks. The study area was subsequently affected by multiple events of uplift and erosion. Four panels out of the 148 modeled time steps are shown.

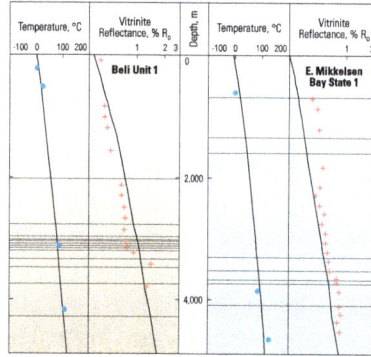

Calibrating modeling results.

Independent information such as subsurface temperature and vitrinite reflectance helps to check the quality of BPSM. Data from two wells show a good fit with model calculations. Recycled vitrinite probably accounts for the elevated reflectance values in rocks shallower than 2,000 m in the E. Mikkelsen Bay State 1 well.

Modeled North Slope accumulations. PetroMod software calculates the paths along which liquid (green) and vapor (red) hydrocarbon phases migrated from areas where they were generated to their accumulation locations.

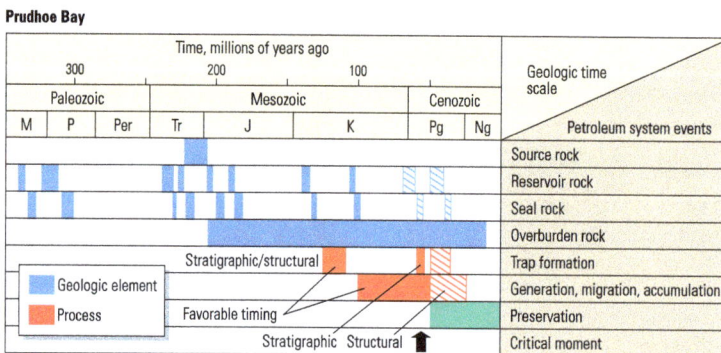

Brooks Range Foothills

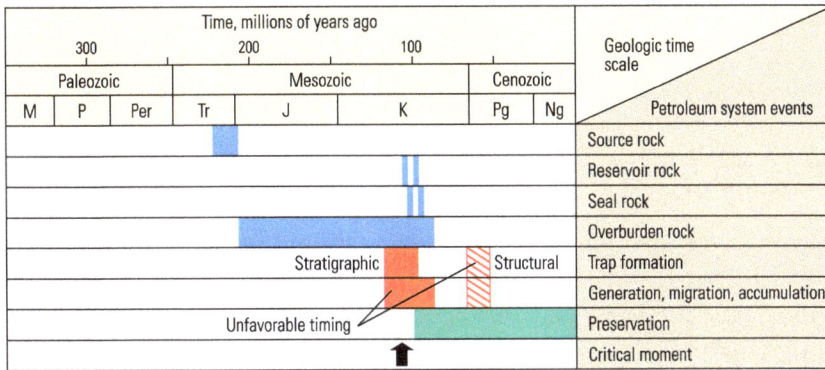

Comparing events charts.

The event chronology for Prudhoe Bay (top) indicates favorable timing for accumulation of hydrocarbons generated from the Shublik source rock. By the time hydrocarbons were migrating in the middle of the Cretaceous (K), many traps had formed and were available to capture fluids. To the south, in the foothills of the Brooks Range (bottom), events were not as favorably timed. Although traps may have formed too late to contain oil and gas generated in the Cretaceous, they might have formed in time to hold remigrating fluids, or those displaced from other areas (hatched).

Simulation results show that hydrocarbon charging occurs quickly—instantaneously on a geologic time scale. If traps are not formed as soon as or before hydrocarbons are ready to move, there is a high risk the fluids will not be trapped. Events charts for two different areas overlying the thermally mature Shublik source rock show how relative timing between trap formation and source-rock maturation can impact risk. At Prudhoe Bay and elsewhere on the Barrow Arch, trap formation preceded generation, migration and accumulation by several million years, resulting in major oil accumulations. However, the events chart at a well in the foothills of the Brooks Range shows that location has significant timing risks for stratigraphic traps, which formed at about the same time as generation and migration of fluids from the Shublik Formation. In addition, risk is high for the structural traps because they can be filled only by remigration of petroleum from older stratigraphic traps.

The various North Slope source rocks matured and expelled petroleum at different times and places, charging reservoirs with a mixture of crude oils. Analysis of biomarkers and stable carbon isotope ratios for oils recovered from wells on the Barrow Arch shows a geographic variation in contributing source rocks. Reservoirs in the west produce oil generated predominately from the Shublik Formation, while those to the east produce oil generated mainly from the Hue-gamma ray zone (Hue-GRZ). The Prudhoe Bay field is intermediate in position and produces oil that is more evenly mixed, containing oil from the Shublik Formation and Hue-GRZ, with lesser input from the Kingak Shale. These findings are consistent with the multiple charging episodes depicted in the 3D PetroMod model, in which the Shublik and Kingak source rocks started to generate and expel petroleum during the Cretaceous, and the Hue-GRZ contributed oil later—and continues to do so today.

Biomarker

Generally, biomarkers are naturally occurring, ubiquitous and stable complexes that are objectively measured and evaluated as an indicator of a certain state. It is used in many scientific fields; medicine, cell biology, exposure assessment, astrobiology, geology and petroleum. Due to the variety of geological conditions and ages under which oil was formed, every crude oil exhibits a unique biomarker fingerprint. Crude oils compositions vary widely depending on the oil sources, the thermal regime during oil generation, the geological migration and the reservoir conditions. Crude oils can have large differences in:

1. Distribution patterns of the n-alkanes, iso-alkanes and cyclic-alkanes as well as the unresolved complex mixture (UCM) profiles

2. Relative ratios of isoprenoid to normal alkanes

3. Distribution patterns and concentrations of alkylated polynuclear aromatic hydrocarbons (PAHs) homologues.

Most of these constituents undergo changes in their chemical structure by time as an effect of several factors among which are the biodegradation and weathering. Relative to other hydrocarbon groups in oil, there are some compounds that are more degradation-resistant in the environment as for example; Pristane, phytane, steranes, triterpanes and porphyrins. These undegradable compounds are known as Biomarkers.

Trebs was the first one to develop the biomarkers concept, with his pioneering work on the identification of porphyrins in crude oils suugesting that these porphyrins are generated from chlorophyll of plants. Blumer et al. and Blumer & Thomas isolated pristane from recent marine sediments and concluded that it was derived from the phytol side chain of chlorophyll. Later, other workers reported the present of various classes of degradationresistant organic compounds and recognized their biomarker implementations.

Petroleum biomarkers can thus be defined as complex organic compounds derived from formerly living organisms found in oil. They show little or no changes in their structure from the parent organic molecules and this distinguishes biomarkers from other compounds. Various biomarkers formed under different geological conditions and ages can occur in different carbon ranges exhibiting different biomarker fingerprints.

From the identification point of view, biomarkers are the most important hydrocarbon groups in petroleum because they can be used for chemical fingerprinting which provides unique clues to the identity of source rocks from which petroleum samples are derived, the biological source organisms which generated the organic matter, the environmental conditions that prevailed in the water column and sediment at the time, the degree of microbial biodegradation and the thermal history (maturity) of both the rock and the oil. The information from biomarker analysis can be used also to determine the migration pathways from a source rock to the reservoir for the correlation of oils in terms of oil-to-oil and oil-to-source rock and the source potential. Also chemical analysis of biomarkers generates information of great importance to environmental forensic investigations in terms of determining the source of spilled oil, differentiating and correlating oils and monitoring the degradation process and weathering state of oils under a wide variety of conditions.

Terpanes and steranes are highly resistant to biodegradation but few studies have shown that they can be degraded to certain degree under severe weathering conditions i.e, extensive microbial degradation.

Biomarkers Analysis

The commercial availability of Gas Chromatography-Mass Spectroscopy (GC-MS) and associated data systems in the mid-1970s led to use the biomarkers for a wide variety purposes. The complex structure of biomarkers and the possible presence in low concentrations make a pressing need for more sensitive and precise analysis. The development of analytical methodologies and the combination between these methods are of great importance to separate, monitor and detect the absolute concentrations and structure of petroleum biomarkers. The use of "hybrid" or "hyphenated" techniques, which are a combination of different separate techniques, increases the analytical power of the used methods. GC-MS can be considered as the most popular method used in the characterization of major biomarker groups. GC provides the significant advantage of the separation of different structures of biomarkers while MS can accurately detect and identify these structures. The concept and the development of these instrumentations will be briefly mentioned.

Separation by Chromatographic Techniques

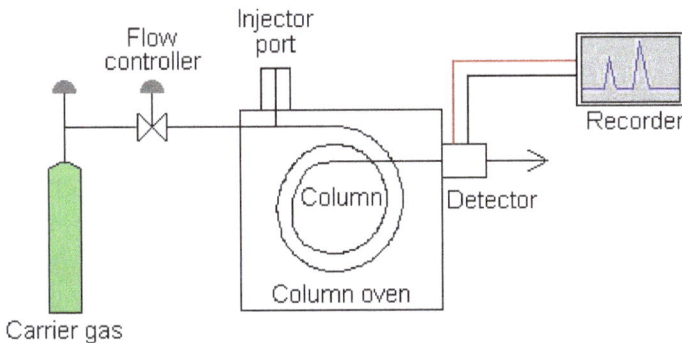

Figure: Schematic diagram of gas chromatograph.

Chromatography is the separation of a mixture of compounds into their individual components primarily according to their volatilities. There are numerous chromatographic techniques but gas chromatography (GC) is the most important one. It has a number of advantages over other separation techniques. It can identify (qualitate) and measure the amount (quantitate) various sample components.

The GC column is the heart of the system; the structure of the stationary phase and the packed material greatly influence the separation of the compounds and affect the time of separation (retention time). Two types, packed and capillary columns, have been used. The advantages in capillary columns over packed columns are in obtaining practically improved resolution in order to give fine structured chromatographic fingerprints. The column is placed in an oven where the temperature can be controlled very accurately over a wide range of temperatures. As compounds come off the column, they enter a detector for identification. Figure below represent the carbon number range distribution of common hydrocarbons in crude oil and petroleum products.

Identification by Spectroscopic Techniques

There are different types of detectors that can be employed depending on the compounds to be analyzed. Mass spectrometry (MS) has very common use in analytical laboratories that study a great variety of compounds and provides a satisfactory tool for obtaining specific fingerprints for classes and homologous series of compounds resolved by gas chromatography. The technique has both qualitative and quantitative uses include identifying and determining the structure of a compound by observing its fragments.

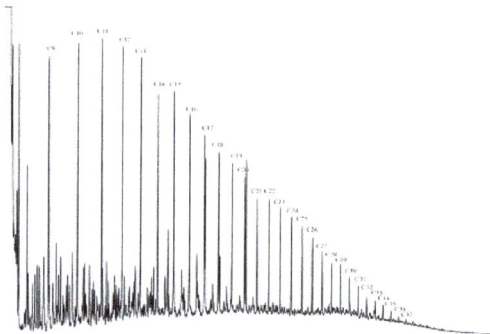

Figure: Representative model of carbon number distribution in petroleum hydrocarbons.

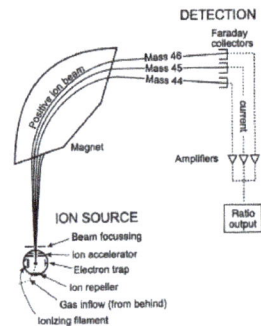

Figure: Schematic diagram of mass spectrometer.

The mass spectrometer has long been recognized as the most powerful detector for gas chromatography. Typical MS instruments consist of three modules; an ion source: which can convert the separated constituent into ions, a mass analyzer: which sorts and separates ions by their masses by applying electromagnetic fields and a detector: which calculate the abundances of each ion present by a quantitative method to generate signals. The size of the signals corresponds to the amount the compound present in the sample.

Characterization of some major biomarker groups is largely achieved using the following MS fragment ions:

- Alkyl-cyclohexanes: m/z 83

- Methyl-alkyl-cyclohexanes: m/z 97

- Isoalkanes and isoprenoids: m/z 113, 127, 183

- Sesquiterpanes: m/z 123

- Adamantanes: m/z 135, 136, 149, 163, 177, and 191

- Diamantanes: m/z 187, 188, 201, 215 and 229

- Tri-, tetra-, Penta-cyclic terpanes: m/z 191

- 25-norhopanes: m/z 177

- 28,30-bisnorhopanes: m/z 163, 191

- Steranes: m/z 217, 218

- 5α (H)-Steranes: m/z 149, 217, 218

- 5β (H)-Steranes: m/z 151, 217, 218

- X diasteranes: m/z 217, 218, 259

- Methyl-steranes: m/z 217, 218, 231, 232

- Monoaromatic steranes: m/z 253

- Triaromatic steranes: m/z 231

The m/z 191 fragment is often the base peak of mass spectra of biomarkers. In general, GC/MS chromatograms of terpanes (m/z 191) are characterized by the terpane distribution in a wide range from C19 to C35 with C29 $\alpha\beta$ - and C30 $\alpha\beta$ -pentacyclic hopanes and C23 and C24 tricyclic terpanes being often the most abundant. As for steranes (at m/z 217 and 218), the dominance of C27,C28 and C29 homologues.

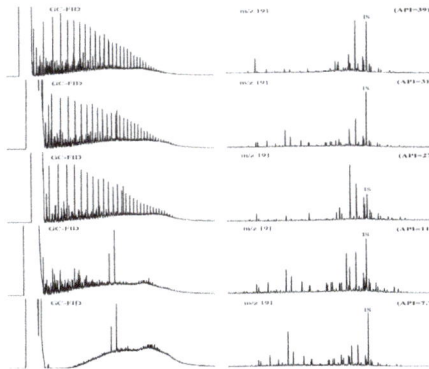

Figure: GC-MS chromatograms at m/z 191 for light to heavy crude oils.

API: American Petroleum Institute. The larger the API, the greater the amount of the light components the oil contains and with decreasing the API, the amounts of medium and heavy weight components increase.

GC-MS chromatograms at m/z 191 for light (API > 35), medium (API: 25–35), and heavy (API < 25) crude oils.

Diagnostic Ratios of Biomarkers

Biomarker diagnostic parameters have been long established and are widely used by geochemists for oil correlation; determination of organic input and precursors, depositional environment, assessment of thermal maturity and evaluation of in-reservoir oil biodegradation. Diagnostic ratios (DRs) can either be calculated from quantitative (i.e., compound concentrations) or semi-quantitative data (i.e., peak areas or heights). Also, many diagnostic ratios currently used in oil spill and environmental studies. Oil–oil correlations are based on the concept that the composition of biomarkers in spill samples does not differ from those of the candidate source oils. Most biomarkers in spill samples and source oils, in particular those homologous series of biomarkers with similar structure, show little or no changes in their diagnostic ratios. An important benefit of comparing diagnostic ratios of spilled oil and suspected source oils is that concentration effects are minimized. In addition, the use of ratios (rather than absolute values) tends to induce a self-normalizing effect on the data because the variations due to the fluctuation of instrument operating conditions day-to-day, operator, and matrix effects are minimized. Therefore, comparison of diagnostic ratios reflects more directly differences of the target biomarker distribution between samples.

Biomarker classes	Diagnostic ratios
Acyclic Isoprenoids	pristane/phytane pristane/n-C17 phytane/n-C18
Terpanes	C21/C23 tricyclic terpane C23/C24 tricyclic terpane C23 tricyclic terpane/C30 $\alpha\beta$ hopane C24 tricyclic terpane/C30 $\alpha\beta$ hopane C24 tetracyclic/C26 tricyclic (S)/C26 tricyclic (R) terpane C2718 α, 21 β -trisnorhopane/C27 17 α, 21 β -trisnorhopane C28 bisnorhopane/C30 $\alpha\beta$ hopane C29 $\alpha\beta$ -25-norhopane/C30 $\alpha\beta$ hopane C29 $\alpha\beta$ -30-norhopane/C30 $\alpha\beta$ hopane oleanane/C30 $\alpha\beta$ hopane moretane(C30 $\beta\alpha$ hopane)/C30 $\alpha\beta$ hopane gammacerane/C30 $\alpha\beta$ hopane tricyclic terpanes (C19-C26)/C30 $\alpha\beta$ hopane C31 homohopane (22S)/C31 homohopane (22R) C32 bishomohopane (22S)/C32 bishomohopane (22R) C33 trishomohopane (22S)/C33 trishomohopane (22R) Relative homohopane distribution (C31-C35)/C30 $\alpha\beta$ hopane homohopane index

Steranes	Relative distribution of regular C27-C28-C29 steranes C27 $\alpha\beta\beta$ /C29 $\alpha\beta\beta$ steranes (at m/z 218) C27 $\alpha\beta\beta$ /(C27$\alpha\beta\beta$ + C28 $\alpha\beta\beta$ + C29$\alpha\beta\beta$) (at m/z 218) C28 $\alpha\beta\beta$ /C29 $\alpha\beta\beta$ steranes (at m/z 218) C28 $\alpha\beta\beta$ /(C27$\alpha\beta\beta$ + C28 $\alpha\beta\beta$ + C29$\alpha\beta\beta$) (at m/z 218) C29 $\alpha\beta\beta$ /(C27$\alpha\beta\beta$ + C28 $\alpha\beta\beta$ + C29$\alpha\beta\beta$) (at m/z 218) C30 sterane index: C30/(C27 to C30) steranes selected diasteranes/regular steranes Regular C27-C28-C29 steranes/C30 $\alpha\beta$ -hopanes
Monoaromatic steranes	C27-C28-C29 monoaromatic steranes (MA) distribution.
Triaromatic steranes	C20 TA/(C20 TA + C21 TA) C26 TA (20S)/sum of C26 TA (20S) through C28 TA (20R) C27 TA (20R)/C28 TA (20R) C28 TA (20R)/C28 TA (20S) C26 TA (20S)/[C26 TA (20S) +C28 TA (20S)] C28 TA (20S)/[C26 TA (20S) +C28 TA (20S)]

Table: Examples of some diagnostic ratios of biomarkers frequently used for the environmental forensic studies.

Examples of Parameters used in Fingerprinting

Normal Alkanes Characteristics

The distribution of n-alkanes in crude oils can be used to indicate the organic matter source. For example, the increase in the n-C15 to n-C20 suggests marine organic matters with contribution to the biomass from algae and plankton. Oil samples characterized by uniformity in n-alkanes distribution patterns suggest that they are related and have undergone similar histories with no signs of biodegradation.

Carbon Preference Index (CPI)

Carbon preference index, obtained from the distribution of n-alkanes, is the ratio obtained by dividing the sum of the odd carbon-numbered alkanes to the sum of the even carbonnumbered alkanes. CPI is affected by both source and maturity of crude oils. CPI of petroleum oils ranging about 1.00 generally shows no even or odd carbon preference indicates mature samples. Also, it can be used in source identification; petroleum origin contaminants characteristically have CPI values close to one.

Degree of Waxiness

The degree of waxiness can be expressed by the ΣC21-C31/ΣC15-C20 ratios. The oils characterized by high abundance of n-C15to n-C20 n-alkanes in the saturate fractions reflecting low waxy. Generally, the degree of waxness < 1 reveals low waxy nature and suggests marine organic sources mainly of higher plants deposited under reducing condition.

Examples of Parameters used in Biomarker Fingerprinting

Pristane/phytane Ratio

Both pristine (2,6,10,14- tetramethyl pentadecane) and phytane (2,6,10,14- tetramethyl hexadecane) are derived from the phytol side chain of chlorophyll, either under reducing conditions (phytane) or oxidizing conditions (pristane). Also both pristine and phytane became dominant saturated hydrocarbon components of highly weathered crude oils until they are degraded.

The pristane/phytane (Pr/Ph) ratio is one of the most commonly used correlation parameters which have been used as an indicator of depositional environment. It is believed to be sensitive to diagenetic conditions; Pr/Ph ratios substantially below unity could be taken as an indicator of petroleum origin and highly reducing depositional environments. Very high Pr/Ph ratios (more than 3) are associated with terrestrial sediments. Pr/Ph ratios ranging between 1 and 3 reflect oxidizing depositional environments.

According to Lijmbach (1975) low Pr/Ph values (<2) indicate aquatic depositional environments including marine, fresh and brackish water (reducing conditions), intermediate values (2–4) indicate fluviomarine and coastal swamp environments, whereas high values (up to 10) are related to peat swamp depositional environments (oxidizing conditions).

Isopreniods/n-alkanes

Waples (1985) stated that by increasing maturity, n-alkanes are generated faster than iosprenoids in contrast to biodegradation. Accordingly, isopreniods/n-alkanes (Pr/n-C17 and Ph/n-C18) ratios provide valuable information on biodegradation, maturation and diagenetic conditions. The early effect of microbial degradation can be monitored by the ratios of biodegradable to the less degradable compounds. Isoprenoid hydrocarbons are generally more resistant to biodegradation than normal alkanes. Thus, the ratio of the pristane to its neighboring n-alkane C17 is provided as a rough indication to the relative state of biodegradation. This ratio decreases as weathering proceeds.

Steranes (m/z 217) Distribution

The distribution of steranes is best studied on GC/MS by monitoring the ion m/z=217 which is a characteristic fragment in the sterane series. It is agreed that the relative amounts of C27-C29 steranes can be used to give indication of source differences. For example, predominance of C28, C29 and C30 steranes indicate an origin of the oils derived mainly from mixed terrestrial and marine organic sources, while oils show slightly low abundance of C28 and C29 and relatively higher concentrations of C27 steranes indicate more input of marine organic source.

Triterpanes (m/z 191) Distribution

Together with steranes, triterpanes belong to the most important petroleum hydrocarbons that retain the characteristic structure of the original biological compounds. Tricyclic, tetracyclics hopanes and other compounds contribute to the terpane fingerprint mass chromatogram (m/z=191) are commonly used to relate oils and source rocks. Mass fragmentogram at m/z=191 can be used to detect triterpanes in the saturate hydrocarbon fraction.

Tricyclic Terpanes

Aquino et al. indicated that tricyclic terpanes are normally associated with marine source. In addition it has been used as a qualitative indicator of maturity. In high mature oils, the tricyclic terpanes is dominated more than in low mature oils.

Homohopanes

The homohopanes (C31 to C34) are believed to be derived from bacteriopolyhopanol of prokaryotic cell membrane. C35 homohopane may be related to extensive bacterial activity in the depositional environment. Homohopane index can be used as an indicator of the associated organic matter type, as it can also be used to evaluate the oxic/anoxic conditions of source during and immediately after deposition of the source sediments. Low C35 homohopanes is an indicator of highly reducing marine conditions during deposition whereas high C35 homohopane concentrations are generally observed in oxidizing water conditions during deposition, consistent with the oxic conditions.

Gammacerane

Gammacerane, originally thought to be as hypersalinity indicator, is associated with both marine and lacustrine environments of increasing salinity.

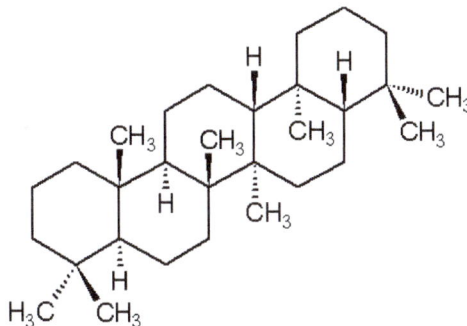

Fig: Gammacerane chemical structure.

Ts/Tm

The ratio of Ts (trisnorneohopane) to Tm (trisnorhopane) more than (0.5) was found to increase as the portion of shale in calcareous facies increases. Van Grass stated that Ts/Tm ratios begin to decrease quite late during maturation but Waples and Machihara reported that Ts/Tm ratio does not appear to be appropriate for quantitative estimation of maturity.

C29/C30 Hopanes Ratios

C29/C30 hopanes ratios are generally high (>1) in oils generated from organic rich carbonates and evaporates.

Steranes/17α (H)-hopanes Ratio

The regular steranes /17 α (H)-hopanes ratio reflects input of eukaryotic (mainly algae and higher plants) versus prokaryotic (bacteria) organisms to the source rock. The sterane/hopane ratio is relatively high in marine organic matter with values generally approaching unity or even higher. In contrast, low steranes and sterane/hopane ratios are more indicative of terrigenous and microbially reworked organic matter.

Bisnorhopanes

It is believed that sediments containing large amounts of bisnorhopane were deposited under anoxic conditions. Bisnorhopanes are types of pentacyclic triterpanes present in significant concentrations in oil. Bisnorhopanes are observed in Guatemalan evaporites and frequency reported in other biogenic siliceous rocks of the circum-Pacific region.

Metalloporphyrins

Porphyrins are the tetrapyrole compounds; the porphyrin nucleus consists of four pyrrole rings joined by four methine bridges giving a cyclic tetrapyrrole structure. The majority of these compounds are thought to originate from various chloropigments produced by phototrophic organisms of the geological past. Metalloporphyrins has become a valuable tool in the determination of the origin and maturity of the organic matter. The porphyrin structure consists of a porphyrin nucleus with various groups of side chains occupying some or all of its peripheral positions.

Metalloporphyrins were extracted from asphaltene and maltene fractions using adsorption column chromatography . Porphyrins occur as etioporphyrin (Etio), Benzo-etio, deoxophylloerythroetioporphyrin (DPEP), Benzo-DPEP and tetrahydrobenzo-DPEP (THBD). The distribution of different types of metalloporphyrins is useful for interpreting transformation of kerogen into bitumen, depositional environments and maturation levels of deposited organic matters.

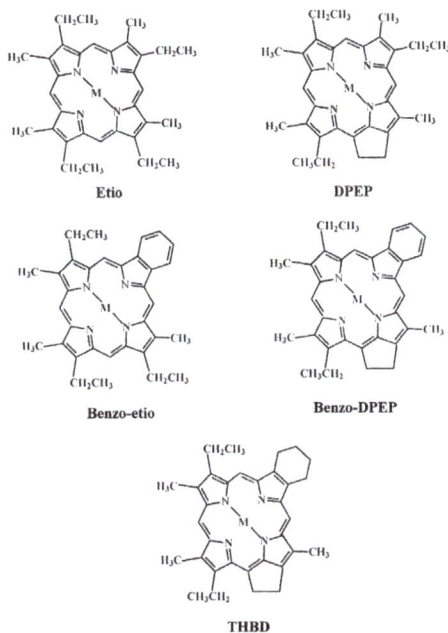

Fig: Structures of different types of metalloporphyrins.

Developments in GC-MS Instrumentation

The low biomarker concentrations in oils (often in the range of several parts per million) in the presence of a highly complex petroleum hydrocarbon matrix especially weathered oils, the variety of chemical classes present in oils and the possible co-elutions in conventional chromatographic separations make the identification of biomarkers a more difficult task.

The development of more reliable, highly selective, fast and sensitive separation and identification tools for biomarker analysis purposes can be considered as one of the most important research points in this field for a meaningful biomarker analysis.

The use of comprehensive two-dimensional gas chromatography (GC × GC) coupled to time-of-flight mass spectrometry (TOFMS) was found to be a powerful tool for overcoming some problems and limitations since it (i) separates substances using two interconnected capillary columns containing different stationary phases and (ii) uses the fast data acquisition of time-of-flight analyzer as a robust registry for GC × GC.

In their work, Aguiar et al. used this technique to overcome the co-elution between tri- and pentacyclic terpanes separated by extracted ion chromatograms (EIC) for ions of mass-to-charge ratio (m/z) 191. The biomarker analysis by GC × GC–TOFMS was much better than in previous works using one-dimensional GC. Co-elutions between tri- and pentacyclic terpanes were clearly resolved in the second column. Noteworthy separation between the C30 hopane and C30 dimethylated homohopane was achieved and overlap of hopanes with steranes in the m/z 217 was eliminated. Besides hopanes, dimethylated triand tetracyclic terpanes were identified. These findings indicate the superiority of GC × GC–TOFMS as a technique for separation and identification of biomarkers in oils due to its high sensitivity, specificity and capability to elucidate compounds structure with high spectral resolution.

Comprehensive two-dimensional gas chromatography (GC×GC) has also been used to separate and identify alkylated aromatics (naphthalenes, biphenyls, fluorenes, phenanthrenes and chrysenes), sulfur-containing aromatics (dibenzothiophenes, benzonaphthothiophenes), steranes, triterpanes, and triaromatic steranes. These biomarkers were separated into easily recognizable bands in the GC×GC chromatogram. Methods used to identify the bands included peak matching with chemical standards and comparison with GC/MS extracted ion chromatograms. By designing mass spectrometers that can determine m/z values accurately to four decimal places, it is possible to distinguish different formulas having the same nominal mass. Since a given nominal mass may correspond to several molecular formulas, lists of such possibilities are especially useful when evaluating the spectrum of an unknown compound.

GC/MS/MS is an operation based on the covariant scan of electrostatic magnetic fields on the trisector double focusing mass spectrometer providing more accurate data. The quadraupole is a common mass separator gives a sufficient sensitivity and selectivity however, high resolution mass spectrometry (HRMS) is also used due to its ability to provide quantitative data for compounds present in complex mixtures for biomarkers analysis. Triple quadrupole GC/MS offers a viable alternation for the rapid, routine analysis providing excellent precision, sensitivity, selectivity, and dynamic range.

Fourier transform ion cyclotron resonance mass spectrometry (FT-ICR MS) benefits from ultra-high mass resolving power (greater than one million), high mass accuracy (less than 1 ppm) and rapid analysis which make it an attractive alternative for the analysis of different and wide range of petroleum products.

It should be noted, however, that there is no single fingerprinting technique that can fully and readily meet the objectives of biomarkers investigation and quantitatively allocate hydrocarbons to their respective sources, particularly for complex hydrocarbon mixtures or extensively weathered and degraded oil residues. Combined and integrated multiple tools are often necessary under such situations.

Data Analysis by Computerized Techniques

Data analysis is an important part of chemical fingerprinting and a broad collection of statistical techniques has been used for evaluation of data.

After separation and identification of biomarkers, principal component analysis PCA, a mathematical procedure, can be used for analyses of chromatograms using a fast and objective procedure with more comprehensive data usage compared to other fingerprinting methods. The discriminative power of PCA can be enhanced by deselecting the most uncertain variables or scaling them according to their uncertainty.

For example, preprocessing of GC-MS chromatograms followed by principal component analysis (PCA) of oil spill samples collected from the coastal environment in the weeks after the Baltic Carrier oil spill and from the tank of the Baltic Carrier (source oil) was carried out. The preprocessing consists of baseline removal by derivatization, normalization, and alignment using correlation optimized warping. The method was applied to chromatograms of m/z 217 (tricyclic and tetracyclic steranes) of oil spill samples and source oils. The four principal components were interpreted as follows: boiling point range (PC1), clay content (PC2), carbon number distribution of sterols in the source rock (PC3), and thermal maturity of the oil (PC4). The method allows for analyses of chromatograms using a fast and objective procedure and with more comprehensive data usage compared to other fingerprinting methods.

References

- Petroleum-geology-8255: petropedia.com, Retrieved 16 March 2018

- Petroleum-systems-and-elements-of-petroleum-geology: connect.spe.org, Retrieved 20 June 2018

- Oil-shale, science: britannica.com, Retrieved 10 July 2018

- What-is-a-petroleum-seep-34142: petro-online.com, Retrieved 27 April 2018

- InTech-Biomarkers-32745: cdn.intechopen.com, Retrieved 26 May 2018

Chapter 7

Petroleum Exploration

Petroleum exploration refers to the search for petroleum deposits beneath the surface of the Earth. It is under the domain of petroleum geology. Various methods are employed in petroleum exploration, such as geological survey, geochemical survey, remote sensing survey, drilling, etc. These have been extensively discussed in this chapter.

We may define petroleum exploration in several ways:

- It is the process of exploring for oil and gas resources in the earth's sedimentary basins. The process relies on the methodical application of technology by creative geoscientists that leads to viable prospects to drill and the actual drilling of these prospects with exploratory and appraisal wells.

- It is the commitment of large amounts of risk capital to explore prospects that have an uncertain outcome.

- It is the primary way in which producing companies—replace their reserves and grow, and the way in which small companies, through a major discovery, may become giants overnight.

- Most of all, it is a necessary core competency for an upstream oil and gas company. If you have the right exploration strategy, capable geoscientists, access to exploration acreage, deep pockets of risk capital and a little luck, you will be successful. If not, you will have modest results—or you may even "bite the dust."

The petroleum exploration process, like the process of buying common stock, involves a series of decisions made under uncertainty—we do not know whether oil or gas is present until after an exploratory well has been drilled. To manage this uncertainty, companies often spread their risk capital over a portfolio of prospects rather than putting all of their investment into one prospect ("putting all their eggs in one basket"). This gives rise to companies sharing prospects among each other through multiple "joint ventures." Sometimes one company is responsible for generating a prospect but, as the capital costs of drilling and field commitment loom, it "sells" partial ownership of the prospect to other companies who "farm in" to the joint venture in return for a drilling commitment. In this way a company may participate in multiple prospects in a given year by holding less than 100% stake in each.

There are some requirements to meet a petroleum reservoir, which is :

1. Source Rock

2. Reservoir Rock

3. Hydrocarbon Trap

Source rock maturity for the petroleum reservoir depends on pressure and temperature during the lithification process, time, and chemical reaction between carbon and hydrogen.

The maturity of the source rock is very important to define the capability of the reservoir can be produced or not. Limestone and shale are the major type of source rocks. Coal may also be considered as a source rock, even though it is present in relatively small quantities within the earth compared to the other source. These source rocks can or can not be a reservoir rock. The concept of reservoir rock is a storage that contain hydrocarbon gas, hydrocarbon liquid, and salt water. The hydrocarbon fluids is less dense than the salt water. So, when it meets salt water, the hydrocarbon fluids will force salt water to migrate downward below. This migration is controlled by rock's permeability and porosity.

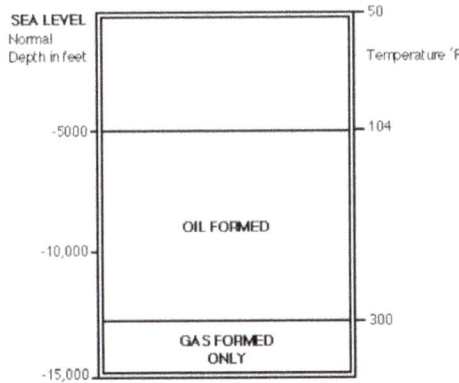

Figure: Temperature Window for Generating Hydrocarbons

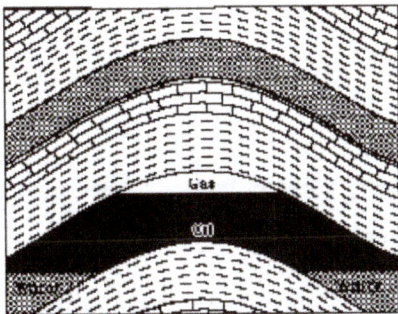

Figure: Hydrocarbon Reservoir with Gas Cap, Oil Zone, and Water Zone

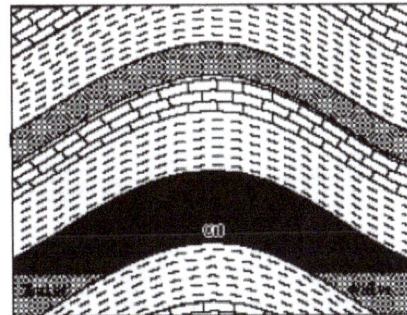

Figure: Hydrocarbon Reservoir with No Gas Cap (Oil Zone and Water Zone Only

During the migration, the reservoir rock needs a trap or seal, so when the fluid migrate it will encounter an impermeable rock. If the fluids continue to migrate when there is no trap or seal, it would go to the surface and dissipate.

Many hydrocarbon reservoirs needs to be mapped before we know the volume and shape of a reservoir. We use petroleum exploration which means process of exploring for oil and gas resources in the earth. How do we explore our huge earth? By some steps, it can be done. Desk study, aerial survey, seismic survey, exploratory drilling, appraisal, development and production are steps to assess a designate field for oil or gas prone.

Desk Study

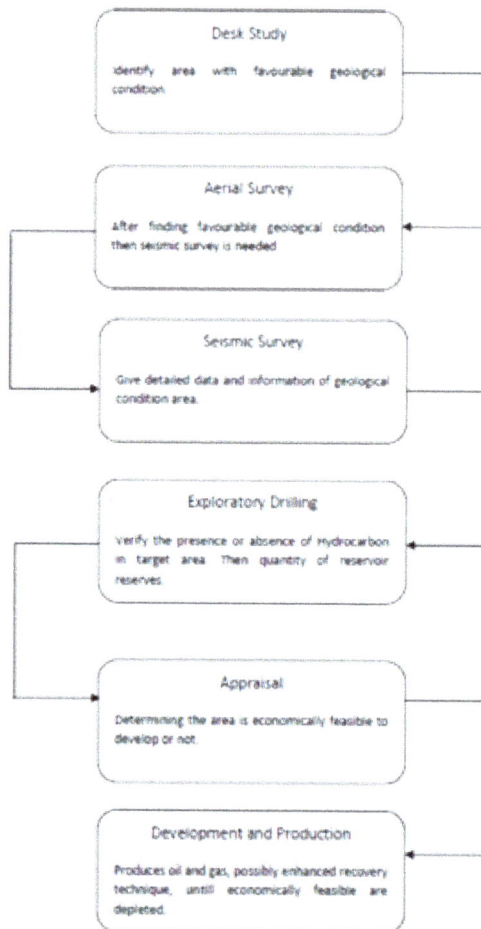

Petroleum Exploration Flowchart

How to identifies an area with a possibility of petroleum system below the surface? Desk study is the basic, for a geologist, to define the geological conditions below the surface. It can be finished by learning the topographic map, geological map, and literature review of certain area. Literature review is the main study research to review the current issues of petroleum exploration and development, planning, and project issues of natural resources and other relevant secondary data information. This secondary data is collected during literature review and during fieldwork interviews by collecting data statistics, documents, or literature about the area. It can be proceeded through a review of academic sources information and discussions.

Aerial Survey

After desk study finds favorable geological condition features that fit for petroleum system, the aerial survey is conducted. Aerial survey is conducted by remote sensing. A geological method that used in order to extract geological information from remote sensing image. A remote sensing images analysis can be carried out with geomorphologic information, structure, and to define rock type classification. Seismic survey will be done when aerial survey information are prospective for a explorationist.

Seismic Survey

In petroleum survey, conducting seismic survey is important to achieve subsurface imagery of survey area. It is operated by contractors (oil field service company) or the oil company itself. Seismic survey can be conducted whether in land, marine, or transition zone; and the procedure of all types of seismic survey is quite the same, when source used in field acquisition is geophone and marine is hydrophone.

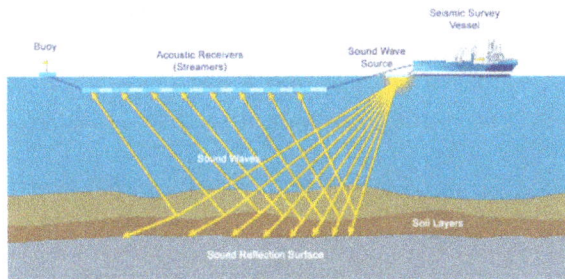

Figure: Marine Seismic Acquisition

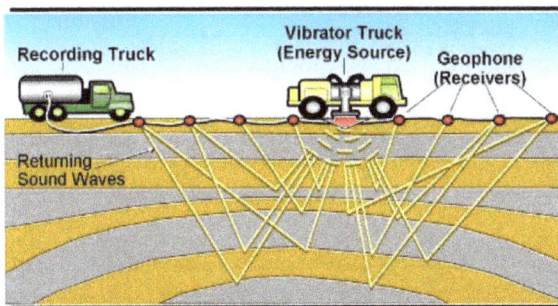

Figure: Land Seismic Acquisition

The survey is started with laying out the line to determine the position of source point and center of geophone group. This work is conducted by the surveyors who also responsible to plan the access route for all survey crew and following works.

In seismic survey, determining field layouts is needed to get expected result. Once the layout is assigned, the rest of survey's procedure must follow the layout plan. The layouts that must be planned include spread type, common-midpoint method, array concept, and other special methods.

Spread is relative location of source and center of geophones used to record reflected seismic wave. Particular type of spread to reduce the noise and unusable trace recorded by geophone group, which lies adjacent with the source. The common spread type used is asymmetrical spread.

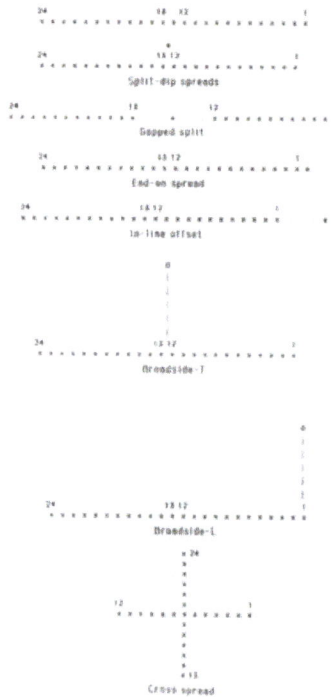

Figure: Types of spread. The o represent source location and x represent geophone group location

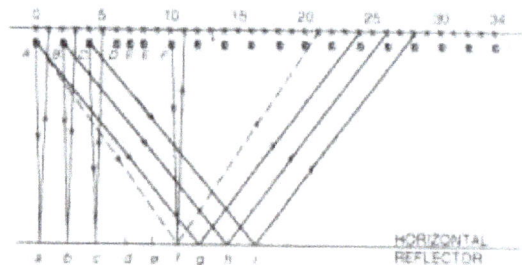

Figure: Vertical section illustrating CMP recording The o represent source location and x represent geophone group location

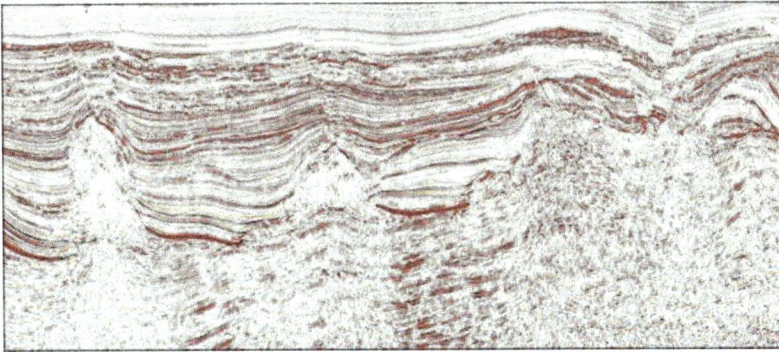

Figure: Seismic section as result of seismic data processing

Common-Midpoint (CMP) methods is used when each reflecting point is sampled than one. Although some surveys conduct single fold recording, CMP method is common in seismic survey. Recorded data can be displayed in common-shot gather, common-receiver gather, or common-midpoint gather.

Array is distribution of sources that fired simultaneously. Arrays will provide a way to discriminate waves which arrived from different directions, either vertically or horizontally from source arrays. Determining array concepts will include tapered arrays, areal arrays, and array constraints. Types of arrays used are sometimes determined from preliminary noise analysis.

After planning field layout, determining field parameters is not less important. Some of field parameters are maximum offset, minimum offset, array length, group interval, and number of channel. Other field parameters sometimes are obtained with field testing. To define the near surface, uphole survey is conducted to find the thickness and velocity of low- velocity layer; which will be followed by seismic refraction.

When the acquisition is complete, data processing is conducted to increase signal to noise ratio and obtain imagery of the subsurface. The process includes noise attenuation, velocity analysis, deconvolution, migration, stacking, and so on. Once the data have been processed and resulted in form of seismic section, the seismic sequence will be interpreted to find out the underlying structure, stratigraphic sequence, and find the petroleum system.

Exploratory Drilling

The data collected from the geologic and geophysical surveys will conclude to a possibility of oil resources location. However, the conclusion is not definite enough before petroleum is now to be existed in the trap, and the quantity is sufficient to make the production activity economical. The only way to provide a definite answers is to drill and test exploratory well(s).

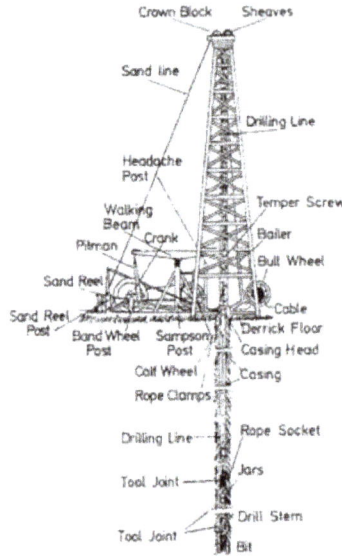

Figure: Standard cable tool drilling system

Walakpa 2

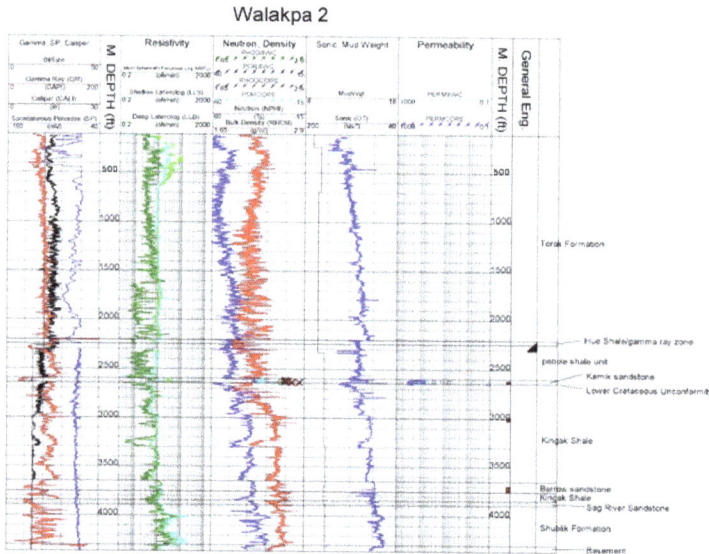

Figure: Well log obtained from wire line logging

The exploratory well location is determined by geologists and geophysicists. Samples of the rock cuttings obtained from exploratory drilling are collected and examined to identify the type of formation and the presence of hydrocarbon materials within the rock based on sample's composition.

Besides rock cutting examination, well logging are also run. Well logging will provide information about rock and fluid properties based on type of logging operated. The tools are placed to the well using electric cable (wireline). There are various type of wireline logging: radioactive log, consists of Gamma Ray, Neutron, and Density; electric log, consists of SP and resistivity; acoustic log; image log; and sampling. Logging will provide information of rock or fluid properties versus depth.

The data provided by one exploratory well may not be sufficient, so additional exploratory drilling is needed to provide better information of the petroleum system. Along with seismic section, well logs will provide data to calculate volume of hydrocarbon stored and conclude whether the production will be economically feasible.

Appraisal

You can interpret your subsurface area after you got the exploratory drilling data. From that, you can predict the petroleum system there. First of all, you have to know is there any source rock in your target or not. If your target doesn't have any source rock, it will absolutely no petroleum system there. After you find the source rock, you need to know is the source rocks mature enough or not. It depends on when formed. If it is not mature enough, it can't generated any oil or expelled it. Then, if there any source rock, we have to look at the other elements of petroleum system. It should has reservoir rock to absorb oil, seal rocks to seal the oil, trap to keep the oil accumulation, and you have to know when and where is the migration path. Afterward, thing that need to know before calculating the resources is, is it oil or gas prone, because oil and gas has their own treatment to maintain the production.

This appraisal is used to know the extent of the discovery. Hydrocarbon reservoir properties, connectivity of petroleum system, hydrocarbon type and gas-oil and oil-water contacts are determined to calculate potential recoverable volumes. This is usually done by drilling more appraisal wells around the initial exploration well. Production tests may also give insight in reservoir pressures and connectivity. And so geochemical and petro physical analysis is also gives more information of the type of the target like its viscosity, chemistry, API, carbon content, etc.

You need to calculate the total reserve of the reservoir. The calculation is given by volumetric method, this is the calculation for oil in place:

$$N(t) = \frac{V_b \phi(p(t))(1 - S_w(t))}{B_o(p(t))}$$

Where

   $N(t)$ = oil in place at time t, STB

   B_b = 7758 A h = bulk reservoir volume, bbl

   7758 = bbl/acre-ft

   A = area, acres

h = thickness, ft

$\varphi(p(t))$ = porosity at reservoir pressure p, fraction

$S_w(t)$ = water saturation at time t, fraction

$B_o(p(t))$ = oil formation volume factor at reservoir pressure p, bb/STB

p(t) = reservoir pressure at time t, psia

And if you find gas, this is the calculation of gas in place:

$$G(t) = \frac{V_b \phi(p(t))(1 - S_w(t))}{(B_g)p(t)}$$

Where

G(t) = gas in place at time t, STB

V_b = 43,560 A h = bulk reservoir volume, ft³

43,560 = ft³/acre-ft

A = area, acres

h = thickness, ft

$\varphi(p(t))$ = porosity at reservoir pressure p, fraction

$S_w(t)$ = water saturation at time t, fraction

$B_g(p(t))$ = gas formation volume factor at reservoir pressure p, ft³/SCF

p(t) = reservoir pressure at time t, psia

Production and Development

Figure: Oil Production Well

Figure: Example of multi-atribute analysis for reservoir

After a hydrocarbon occurrence has been discovered and appraisal has indicated it is economically feasible enough to develop. You can focuses on extracting hydrocarbon. Production wells are drilled and prepared for producing the oil. For optimal recovery 3D seismic is usually available and it can mapped the subsurface area. After prodruting the oil, we need to prepare for enhanced recovery like steam injection, pumps, or any chemical way. This enhanced recovery can extract more hydrocarbon so it can redevelop fields.

Mostly, well is conducted to take oil from reservoir rocks. A reservoir rock like sandstone and fractured limestone can determined through a combination of regional studies. And it can determined from analysis of other wells in target area., stratigraphy and sedimentology, and seismic interpretation. Once a possible hydrocarbon reservoir is identified, the key physical characteristics of a reservoir that are of interest to a hydrocarbon explorationist are its bulk rock volume, net-to-gross ratio, porosity and permeability.

Bulk volume of the rock is total volume of rock above any oil-water contact or gas- water contact, can determined by mapping and correlating sedimentary packages. The net-to- gross ratio, typically estimated from analogues and wireline logs, is used to calculate the proportion of the sedimentary packages that contains reservoir rocks. The bulk rock volume multiplied by the net-to-gross ratio gives the net rock volume of the reservoir. The net rock volume multiplied by porosity gives the total hydrocarbon pore volume i.e. the volume within the sedimentary package that fluids (importantly, hydrocarbons and water) can occupy. The summation of these volumes for a given exploration prospect will allow explorers and commercial analysts to determine whether a prospect is financially viable.

Porosity and permeability were determined through the study of drilling samples, analysis of cores obtained from the wellbore, examination of contiguous parts of the reservoir that outcrop at the surface and by the technique of log analysis using GR, SP, resistivity, or neutron log, that integrated each other. Modern advances in siesmic data acquisition and processing have meant that seismic attributes of subsurface rocks are available and can be used to infer physical/sedimentary properties of the rocks.

Geological Survey

A geological survey is the basic professional work normally done by the geologists. A geological survey is a systematic investigation of the geology of an area. It reflects the geology and structure beneath a given piece of land. Surveys are conducted for the purpose of preparing a geological map. Any geological surveying method employs several techniques including the traditional traverses, walk-over surveys, studying the exposures, outcrops and landforms. Geological surveys also adopt some intrusive methods, like hand augering and machine drilled boreholes. Geological surveys also use the geophysical techniques and remote sensing methods, such as aerial photography and satellite imagery.

Geological surveys are normally undertaken by private agencies, state government departs of mines and geology, and national geological survey organizations. They maintain the geological inventory of various formations, mineral deposits and resources. They keep all records for the advancement of knowledge of geosciences for the benefit of the nation. Geological mapping are parts of a geological survey. It involves certain procedures.

Kinds of Geological Surveys: A geological survey can be undertaken using a number of methods depending on the size of a region and the amount of information that are required. Different types of methods are involved in ecological surveys. The first method is Remote Sensing. It is used in some geological mapping works. This is done using satellite remote sensing methods. While most of these methods rely on geophysical rather than pure geological data, the use of this method can give a broad scale view of surface geological structures such as folding, faulting, igneous intrusions etc. The next method is the Air photo interpretation method. This can give a broad overview of the geological relationships of an area with no detailed knowledge of the mineral composition or fabric of the rocks. The third method is the Outcrop surveying method. This is normally done by geologists by conducting traverses along the fields and mapping the outcropping rock types. The last part is the Geology interpretation surveys. These are more detailed outcrop surveys, where the geological boundaries are established and interpreted in a small area.

Geological mapping and prospecting are valuable techniques in an petroleum exploration. Geological prospecting and exploration for oil and gas is a set of industrial and R&D activities for geological study of subsurface resources, identification of promising areas, discovery of fields, their evaluation and pre-development. The final objective of geological prospecting is preparation of subsurface resources. The main principle of geological prospecting is the comprehensive geological study of subsurface resources when along with oil and gas exploration all associated components (petroleum gas and its composition, sulfur, rare metals, etc.), possibility and practicality of their production or utilization are investigated; hydrogeological, mining, engineering, geological and other studies are performed; natural, climatic, socioeconomic, geological and economic conditions and their changes caused by future field development are analyzed.

Objectives of Geological Field Mapping

There are several reasons based on which a geological field mapping is carried out. They are all entailed in collecting variable amounts of field data. These basic reason is to delineate the natural mineral and other resources. Mineral and oil exploration proceeds always in this way. Geological mapping is usually the first task in any reconnaissance study. Geophysical investigations are carried out to answer the question of the extent of the system under the subsurface. Geochemical investigations are also used to estimate parameters such as the temperature of the system. Exploitation of all mineral resources requires the appreciation of basic geology and optimum utilization of a potential area. This requires the mapping of the resource. In addition, the understanding of the spatial distribution and deformation of rock units, at the surface, is critical in order to develop a 3-dimensional model of the subsurface geology.

Geological mapping: Geological mapping is done to obtain and provide basic knowledge about the prevailing field conditions, not only through direct observations but also by collecting and analyzing rock, mineral and sediment samples. A geologist conducts field surveys and prepare accurate geological maps by collecting samples and measuring the geometrical aspect of outcrops. There is no substitute for a geological map. Geological mapping is normally done in a project mode with people in a team, a set of special equipment, and a topographic base map. Careful observations are done during the geological mapping.

Remote Sensing for Geological mapping: Today, the availability of aerial photography and remote sensing from satellite imagery, and the computer capability for storage, recovery, and evaluation of data are used for geologic mapping and other purposes. These methods have almost replaced many old methods of geologic data collection, plotting, and interpretation. The Remote sensing technology and satellite products provide the fast access to all geospatial data. Also, a greater and finer resolution of data and images are readily available in planimetric and 3-D mode at any desired scale and time. These data can be integrated with Geographic Information Systems (GIS) for vertical and horizontal comparison. Maps can be combined with layers of information on topography, minerals, water, energy, and the environment. These technological advances have increased the usefulness of and public access to geologic maps.

Geological observations: The three basic reasons why geological field work is carried out include exploitation of natural resources, as a requirement of the government and for academic purposes. Good geological mapping should be executed in three phases; planning, data collection and reporting. The data collection phase involves detailed observations. All geological observations are marked on the base maps for future compilation and interpretation. Base maps are used to locate the positions of people objects and structures in the field.

Geochemical Methods: Geochemical methods involve the measurement of the chemistry of the rock, soil, stream sediments or plants to determine abnormal chemical patterns, which may point to areas of mineralization. When a mineral deposit forms, the concentration of the ore "metals" and a number of other elements in the surrounding rocks is usually higher than normal. These patterns are known as primary chemical halos. When a mineral deposit is exposed to surface processes, such as weathering and erosion, these elements become further distributed in the soil, groundwater, stream sediments or plants and this pattern is called a secondary chemical halo. Secondary halos aid in the search for deposits as they normally cover a greater area and therefore 3 the chance of a chemical survey selecting a sample from these areas is greater than from a primary halo area. Different elements have different "mobility" in the environment based on their readiness to dissolve in water, their density, their ability to form compounds with other elements and the acidity (pH) of the environment. Subsequently, the secondary halo may not contain the "metal" for which a geochemical survey is searching but other "marker" elements. These are commonly employed during geological surveys.

Geophysical surveys: Geophysical prospecting method are employed when there are no exposures and the entire region is covered with soils and regoliths. These are indirect method of finding out the hidden rock types and structures.

Geophysical survey refers to the systematic collection of geophysical data for geospatial studies. Geophysical surveys are conducted using a great variety of sensing instruments. The data are collected from above or below the Earth's surface. Sometimes the data are collected from aerial, orbital, or marine platforms. Geophysical surveys have many applications in Earth science, archaeology, mineral and energy exploration, oceanography, and engineering. Geophysical surveys are part of the geological surveys.

Base maps and other Maps: During the preliminary phase, all existing data and maps of the area of study are to be collected and analyzed. All suitable maps available like the physical, political, relief, road, physical, and topographic maps are to be seen first. These are to be carried to the field as it is possible that details in one may not be present in another. Most importantly, for a geological fieldwork, a handy base map is expected to be used as a reference. Depending on the areal extent of the field and the detail required, the scale of the map is to be chosen as it is an important aspect to be considered.

Geological investigations: Geological investigations normally start with base maps, run through the field areas and end in laboratory analysis of samples. The ultimate aim is to explain the geology and structure of the area. The approach is highly practical. The first geologic map was prepared, in the world, to solve a practical problem involving the distribution of different types of rocks at and near the Earth's surface. Most building materials, except wood, are from various specific rocks and rock products. Geological investigations are the basic needs for a country. A geologic map graphically communicates important information about the distribution of rocks and unconsolidated materials at and near the Earth's surface.

Geochemical Survey

Geochemical Exploration Methods are based on the assumption that the hydrocarbon found in an oil pool tent to migrate upwards because of their lower density, some of these hydrocarbon molecules may eventually reach the surface. In the proved oil/gas fields, the samples of surface are likely to have a comparatively high percentage of hydrocarbon content. Similarly, higher than average chloride content could be expected around the edges of an oil pool left by the water which has migrated and evaporated.

Geochemical method is still in an experimental stage and requires extremely precise analysis technique. It is interesting for an oil explorer because of its direct approach. The geochemical methods generally used are:-

a) Micro gas survey: These surveys are prominently carried out in Russia and adjoining countries.

Method: The area under investigation is divided into profiles. The interval of the profiles is decided, depending upon the work and generally the distance of 2 kms taken.

The laying of the profiles is just the same as is being done in case of seismic survey. The samples of the circulating muds were collected at 5 to 10m interval from exploratory wells. After the samples were collected they were taken to the laboratory for degassing. The quantity of sample collected was half a litre. The chromatograph (figure) was used for analyzing the different samples. The quantity of different gas is calculated with the help of the formula:

$$Q = \frac{L.K.V.1.88}{100}$$

Where L= length of the peak

      K =derived value

      V = volume of gas

For conducting the micro gas survey, the following precautions are necessary:

The water used for drilling of the holes should be as pure as possible.

- The bottles in which the samples are collected should be clean and washed with warm water.

- The people conducting the analysis of gas in the laboratory should be so done that they are free from the atmospheric gases.

- Generally the collection of the gas samples should be so done that they are free from the atmospheric gases.

Micro Gas surveys can be used in conjunction with seismic survey to improve the quality of work in a big area and to delineate structures.

The greatest utility of Gas Surveys has been proved in the following circumstances:

- Prospecting for pools in stratigraphic and lithological traps.

- For deciding whether a previously discovered structural trap does contain oil or simply 'dry'.

Various forms of gas surveys have been tried in other countries of the world. On the basis of these trials, it has been confirmed that the best and least ambiguous results are obtained in 'Deep gas survey' in which the subsoil atmosphere from depth 2.60 meters at various levels is examined for hydrocarbon content. A more effective version of the same is 'gas logging' undertaken in shallow, structural and deep exploratory wells. Gas surveys on experimental basis should be taken up in one particular area first and then extended to other regions.

(b) Gas Core Surveys: In some of the areas in the western world, gas core surveys were earlier done. The surveys were done by degassing the cores collected during drilling and analyzing gas. Such surveys are now being discontinued.

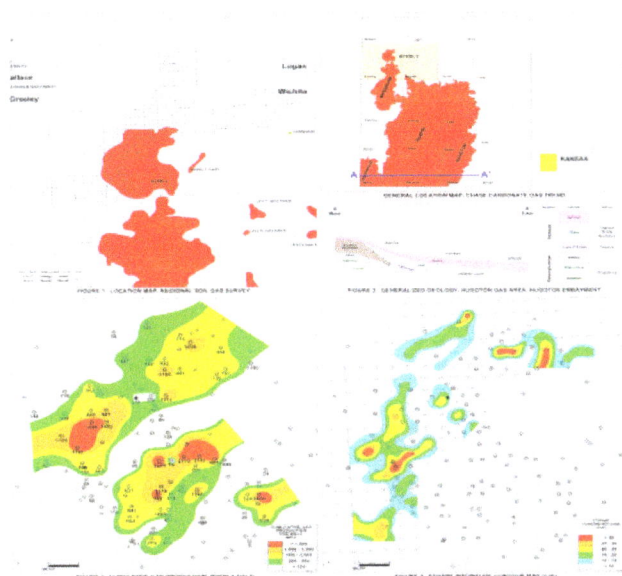

Fig: Soil Gas Survey

(c) Gas Logging: Gas Logging which is one of the geochemical methods of prospecting for hydrocarbons is a continuous method unlike other geophysical methods. Therefore, on exploratory wells where there may be a danger of blow out, application of gas logging is a boon.

Much earlier, to the penetration of productive horizons it can prewar about the approaching pool since it has got the advantage of recording the gas shows which area present in the mud in the form of micro concentrations. Much earlier, to penetration of productive horizons it can pre-warn about the approaching pool since it has got the advantage of recording the gas shows coming through diffusion process in the mud. Till such time, the mud do not heavily enriched with gas, and its parameters do not change appreciably. Therefore, gas logging is the only means on which can be relied for information of the approaching danger of a blow out. With the latest models of gas logging units, one can have on the spot and at the very moment of drilling, an idea about the quantitative and qualitative nature of the hydrocarbon coming from the pool.

(d) Hydro chemical surveys: Another geo chemical method of prospecting for oil/gas is by hydro chemical surveys. As the name suggests, this method analyses formation water properties as it is closely involved in the primary mechanism that causes the accumulation, preservation and destruction of oil and gas fields. Water serves as a vehicle in transporting the hydrocarbons from their source bed to a trap, where they accumulate. Knowledge of the types, class and characteristics of water associated with oil and gas accumulations is needed in geochemical exploration.

Certain constituents dissolved in oil field water are called favorable indicators of hydrocarbon accumulations. Iodide, ammonium salt, organic acid, salts, ethane butane, low sulphate concentrations and the type and class of brine are important. Amount of aromatic hydrocarbon in formation water directly reflects the occurrence of petroleum and can be used to estimate its proximity.

Hitchon and Horn used a statistical technique discriminant analysis to show that formation waters are associated with large hydrocarbon accumulations. According to them iodide and magnesium were important discriminators in water from Paleozoic age rocks with that of water from Mesozoic age rocks.

As per Suilin's classification the main subdivisions of water will be water will be:

(1) Sodium sulphate type   - $Na_2 SO_4$

(2) Sodium bicarbonate - $Na_2 HCO_3$

(3) Magnesium chloride – $MgCl_2$

The oil field waters are highly concentrated in chloride, normally sulphated and saturated with calcium sulphate and carbonate. They generally contain more than 1 mg/lit of iodide and 300 mg/lit of bromide with Cl/Br ratios less than 350 and SO4 x 100 Cl ratios less than 1. Iodide and Bromide are related to bituminous substances and thus to hydro carbon accumulations.

Other important characteristics of oil field water are presence of Benzene (20 mg/lit.), the negative redox potential pressure greater than .65 psi/ft of depth and temperature greater than 66 degree Celsius but less than 149 degree Celsius.

The figure x would show the genetic indicators related to water associated with a reservoir likely to contain oil or gas, while the other one should show the genetic indicators related to water associated with a reservoir not likely to contain accumulations of oil and gas. Appropriate mapping of these indicators with geophysical geological information will prove useful in locating oil and gas pools.

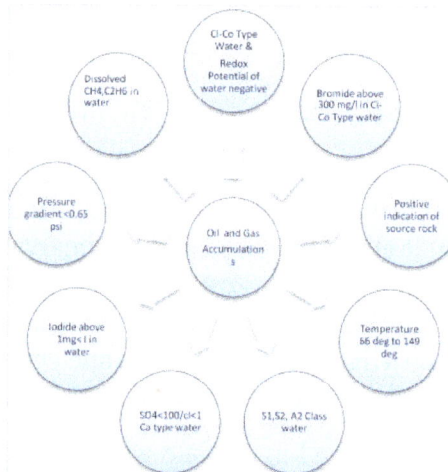

Fig: Oil Detection Indicators

(e) Organo - Hydrochemical Survey: Natural gases are understood to diffuse into edge waters only a few kilometers from petroleum accumulation. Aquifers over-lying oil/gas pools show anomalous concentration of gaseous hydrocarbons.

Dissolved organics have significantly high concentration in interstitial waters in source rocks during primary migration and in water expelled during clay mineral digenesis. Also aerobic oxidation water washing and microbial degradation leave organo-geochemical imprints.. The most commonly employed hydro chemical indices are (i) dissolved gas (ii) total dissolved organic carbon (iii) benzene (iv) naphthenic acids & (v) aliphatic acids.

(f) Asphaltenes as frontier molecules in geochemical research: Asphaltenes are emerging as frontier molecules in geochemical research as they provide important clues about nature of source organics, maturation migration and secondary alteration effects. It is a major class of petroleum components that has no analogous counterpart in the in the biological system. This suggests that petroleum asphaltenes are secondary products formed after the decomposition of source material. Asphaltenes may be fragments of original kerogen from which petroleum is derived and may be expelled as a part of oil.

(g) Sniffer Surveys: The detection of oil and gas seeps in the offshore area by method of sniffing is now being conducted on a routine basis and is one of the most important geochemical methods being followed in the offshore areas. Hydrocarbons seeping from the sea floor dissolve in the sea water and form plumes which are transported by marine currents and mixing. These plumes can be detected at a distance of 10 km from their source area and are sometimes detectable as far as 20 km.

In order to optimize the probability of detecting seep continuous sampling must be conducted at a depth below thermocline. Analytical sensitivity must be of the order of 5×10^{-9} ml gas per ml water. In order to find out the place from where the seep is originating and to simplify data interpretation, the response time of the analytical system must be short. A computer analysis of a mathematical model has been developed to determine optimum lane spacing.

This system provides for direct detection and is specific for hydrocarbons. It would be economical if conducted abroad a vessel on conventional geophysical survey. The sniffer data has been collected and used both in the time of reconnaissance survey and during detailed exploration survey. At the time of reconnaissance, it is done with the conventional seismic equipment. The data obtained with the help of sniffer survey is utilized to confirm that the area of interest is a petroliferous area. Seeps observed are plotted on large scale map and their areal distribution and relationship to subsurface geology and structure are noted.

When this is done along with other geophysical surveys, the sniffer operator makes notes of anomalous areas which should be given special attention during the data interpretation phase. When deployed on a survey where the sniffer is considered to be particularly important the operator make certain operational decisions based on the data at the movement. The sniffer data consists both of hydrographic parameters which effect the distribution of hydrocarbon anomalies and hydrocarbon concentrations. The vessel carrying sniffer survey equipment provides a continuous measurement of the salinity and the temperature and depth sensors located in the towed body. The continuous values of each of the Hydrocarbon components are also plotted.

Remote Sensing Survey

Remote sensing techniques in the exploration for petroleum have not moved from the small-scale, limited-study-area, experimental state to full-scale, large-area, operational status. Remote sensing techniques will have come to maturity when total basin surveys for known and potential hydrocarbon anomalies are common-place. As with much of petroleum exploration, remote sensing is primarily an indirect technique limited to the development of drillable petroleum prospects. Remote sensing techniques include spectroscopic analysis, which offers the potential for airborne geochemical surveys. Research toward the latter objective is still in early phases.

The most commonly used wavelengths are the visible part of the spectrum (0.3-0.7 μ), infrared film emulsions (0.3-1.1μ), and thermal infrared (8-14μ). Equipment and materials covering these spectral bands are the best developed and the most widely available.

Exploration in areas of consistently poor illumination because of meteorologic conditions will bring about increased use of the longer wavelength (microwave) equipment. Cloud penetration is a function of wavelength; passive microwave radiometers, side-looking radar, and scatterometers possess this capability. Currently, airborne microwave instrumentation is not widely available, but indications are that it will come into wider use.

Remote Sensing is a critical element for an effective response to marine oil spills. Timely response to an oil spill requires rapid reconnaissance of the spill site to determine its exact location, extent of oil contamination (particularly the thickest portion of the slick) and verifying predictions of the movement and fate of oil slicks at sea. This is necessary to effectively direct spill countermeasures such as mechanical containment and recovery, dispersant application and in situ burning, the timely protection of sites along threatened coastlines and the preparation of resources for shoreline clean-up. Remote sensing is useful in several modes of oil spill control, including large area surveillance, site specific monitoring and tactical assistance in emergencies. It is able to provide essential information to enhance strategic and tactical decision-making, decreasing response costs by facilitating rapid oil recovery and ultimately minimizing impacts. For ocean spills, remote sensing data can provide information on the rate and direction of oil movement through multi-temporal imaging and input to drift prediction modeling. Observation can be undertaken visually or by use of remote sensing systems. In remote sensing, a sensor other than human vision or conventional photography is used to detect or map oil spills. Remote sensing of oil on land is particularly limited.

Visual observations of spilled oil from the air, along with still and video photography, are the simplest and most common method of determining the location and extent (scale) of an oil spill. Remote sensing of spilled oil can be undertaken by helicopter, particularly over near-shore waters where their flexibility is an advantage along intricate coastline with cliffs, coves and islands. For open ocean spills, there is less need for rapid changes in flying speed, direction and altitude, in these instances the use of low altitude, fixed-wing aircraft have proven to be the most effective tactical method for obtaining information about spills and assisting in spill response. For spill response efforts to be focused on the most significant areas of the spill, it is important to note the relative and heaviest concentrations of oil. GPS and other aircraft positioning systems allow pinpointing the oil's location. Photography, particularly digital photography, is also a useful recording tool and allows others to view the situation on return to base. Many devices employing the visible spectrum, including the conventional video camera, are available at a reasonable cost. Dedicated remote sensing aircraft often have built-in downward looking cameras linked with a GPS to assign accurate geographic co-ordinates.

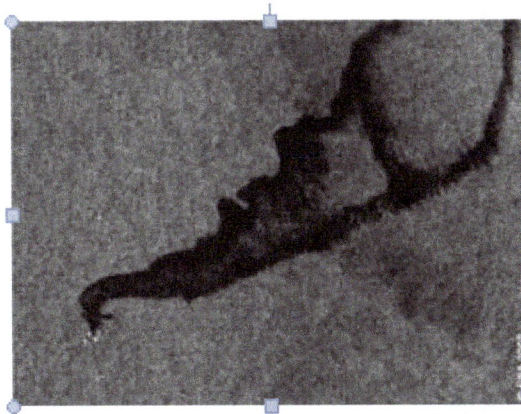

Figure: A satellite image of the same November 20, 2004 event.

Practical oil spill detection is still performed by visual observation, which is limited to favorable sea and atmospheric conditions and is inoperable in rain, fog, or darkness. Visual observations are restricted to documentation of the spill because there is no mechanism for positive oil detection. Very thin oil sheens are also difficult to detect especially in misty or other conditions that limit vision. Oil can be difficult to see in high seas and among debris or weeds where it can blend in to dark backgrounds such as water, soil, or shorelines. Many naturally occurring substances or phenomena can be mistaken for spilled oil. These include sun glint, wind shadows and wind sheens, biogenic or natural oils from fish and plants, glacial flour (finely, ground mineral material usually from glaciers), and oceanic or riverine fronts where two different bodies of water meet. The usefulness of visual observations is limited, however, it is an economical way to document spills and provide baseline data on the extent and movement of the spilled oil

An estimate of the quantity of oil observed at sea is crucial. Observers are generally able to distinguish between sheen and thicker patches of oil. However gauging the oil thickness and coverage is rarely easy and is made more difficult if the sea is rough. All such estimates should be viewed with considerable caution. The table below gives some guidance. Most difficult to assess are water-in-oil emulsions and viscous oils like heavy crude and fuel oil, which can vary in thickness from millimeters to several centimeters.

Oil Type	Appearance	Approximate Thickness	Approximate Volume (m^3/km^2)
Oil Sheen	Silver	>0.0001 mm	0.1
Oil Sheen	Iridescent (rainbow)	>0.0003 mm	0.3
Crude and Fuel Oil	Brown to Black	>0.1 mm	100
Water-in-oil Emulsions	Brown/Orange	>1 mm	1000

Remote sensing equipment mounted in aircraft is increasingly being used to monitor, detect and identify sources of illegal marine discharges and to monitor accidental oil spills. Remote sensing devices used include the use of infra-red (IR) video and photography from airborne platforms, thermal infrared imaging, airborne laser fluourosensors, airborne and satellite optical sensors, as well as airborne and satellite Synthetic Aperture Radar (SAR). SAR sensors have an advantage over optical sensors in that they can provide data under poor weather conditions and during darkness. Remote sensors work by detecting properties of the sea surface: color, reflectance, temperature or roughness. Oil can be detected on the water surface when it modifies one or more of these properties. Cameras relying on visible light are widely used, and may be supplemented by airborne sensors which detect oil outside the visible spectrum and are thus able to provide additional information about the oil. The most commonly employed combinations of sensors include Side-Looking Airborne Radar (SLAR) and downward-looking thermal IR and ultra-violet (UV) detectors or imaging systems. All sensors must be calibrated and require highly trained personnel to operate them and interpret the results.

Satellite-based remote sensing systems can also detect oil on water. The sensors on board are either optical, detecting in the visible and near IR regions of the spectrum, or use radar. Optical observation of spilled oil by satellite requires clear skies, thereby severely limiting the usefulness of such systems. SAR is not restricted by the presence of cloud and is a more useful tool. However, with radar imagery, it is often difficult to be certain that an anomalous feature on a satellite image is caused by the presence of oil. Consequently, radar imagery from SAR requires expert interpretation by suitably trained personnel to avoid other features being mistaken for oil spills. To date, operational use of satellite imagery for oil spill response has not been possible because limited spatial resolution, slow revisit times, and often long delays in receipt of processed image. However, satellite imagery can be used later to complement aerial observations and provide a wider picture of the extent of pollution.

The present inability to reliably detect and map oil trapped in, under, on, or among ice is a critical deficiency, affecting all aspects of response to oil spills in ice. Although there is still no practical operational system to remotely detect or map oil-in-ice, there are several technology areas where further research into ground-based remote sensing could yield major benefits. Examples include ground-penetrating radar, optical beams for river spills and vapor detection (e.g. gas-sniffer systems) for oil trapped in and under ice.

A critical gap in responding to oil spills is the present lack of capability to measure and accurately map the thickness of spilled oil on the water. There are no operational sensors, currently available, that provide absolute measurement of oil slick thickness on the surface of the water. A thickness sensor would allow spill countermeasures to be effectively directed to the thickest portions of the oil slick. Some IR sensors have the ability to measure relative oil thickness. Thick oil appears hotter than the surrounding water during daytime. Composite images of an oil slick in both UV and IR sensors have shown able to show relative thickness in various areas with the thicker portions mapped in IR and the thin portions mapped in UV.

So the remote sensing technique helps to monitor the change of the location for the oil spill, measure the distance, time, the change in size and shape.

Geophysical Methods

Surface geophysical techniques determine density, magnetic, and acoustical properties of a geologic medium. Three geophysical methods used in petroleum exploration comprise magnetic, gravimetric, and seismic (including refraction/reflection) techniques. The magnetic and gravity methods are used only in primary surveys where little is known of the subsurface geology and the thickness of sediments of potential prospective interest. The seismic reflection method is universally used for determining the underground geological structure of a reservoir rock in a certain area. The method(s) selected will depend on the type of information needed, the nature of the subsurface materials, and the cultural interference.

Gravity Methods

The gravity field of the Earth can be measured by timing the free fall of an object in a vacuum, by measuring the period of a pendulum, or in various other ways. Today almost all gravity surveying is done with gravimeters. Such an instrument typically consists of a weight attached to a spring that stretches or contracts corresponding to an increase or decrease in gravity. It is designed to measure differences in gravity accelerations rather than absolute magnitudes. Gravimeters used in geophysical surveys have an accuracy of about 0.01 milligal (mgal; 1 mgal = 0.001 centimeter per second per second). That is to say, they are capable of detecting differences in the Earth's gravitational field as small as one part in 100,000,000.

Gravity differences over the earth's surface occur because of local density differences between adjacent rocks. The variations in the density of the crust and cover are presented on a *gravity anomaly map*. A gravity anomaly map looks at the difference between the value of gravity measured at a particular place and the predicted value for that place. Gravity anomalies form a pattern, which may be mapped as an image or by contours. The wavelength and amplitude of the gravity anomalies gives geoscientists an idea of the size and depth of the geological structures causing these anomalies. Deposits of very dense and heavy minerals will also affect gravity at a given point and produce an anomaly above normal background levels.

Anomalies of exploration interest are often about 0.2 mgal. Data have to be corrected for variations due to elevation (one meter is equivalent to about 0.2 mgal), latitude (100 meters are equivalent to about 0.08 mgal), and other factors. Gravity surveys on land often involve meter readings every kilometer along traverse loops a few kilometers across. It takes only a few minutes to read a gravimeter, but determining location and elevation accurately requires much effort.

Gravity measurements can be obtained either from airborne (remote) or ground surveys. The most sensitive surveys are currently achieved from the ground. Variations of gravity are due to local changes in rock density and therefore depend on the type of rocks beneath the surface. Sedimentary rocks are, for example, less dense than granite, which is in turn less dense than basalt.

> High Density
>
> > Extrusive Igneous Rocks, e.g. Basalt
> >
> > Metamorphic Rocks
> >
> > Intrusive Igneous Rocks, e.g. Granite
> >
> > Sedimentary Rocks
>
> Low Density

In most cases, the density of sedimentary rocks increases with depth because increasing pressure reduces porosity. Uplifts usually bring denser rocks nearer the surface and thereby create positive gravity anomalies. Faults that displace rocks of different densities also can cause gravity anomalies. Salt domes generally produce negative anomalies because salt is less dense than the surrounding rocks. Such faults, folds, and salt domes trap oil, and so the detection of gravity anomalies associated with them are crucial in petroleum exploration. Moreover, gravity measurements are occasionally used to evaluate the amount of high-density mineral present in an ore body. They also provide a means of locating hidden caverns, old mine workings, and other subterranean cavities.

Density contrasts of different materials are also controlled by a number of other factors. The most important are the grain density of the particles forming the material, the porosity of the material, and the interstitial fluids within the material. Generally, specific gravities of soil and shale range from 1.7 to 2.2. Massive limestone averages 2.7. While this range of values may appear to be fairly large, local contrasts will be only a fraction of this range. A common order of magnitude for local density contrasts is 0.25.

Gravity surveys provide an inexpensive method of determining regional structures that may be associated with groundwater aquifers or petroleum traps. Gravity surveys have been one of the principal exploration tools in regional petroleum exploration surveys. Gravity surveys have somewhat limited applications in geotechnical investigations.

Electrical Methods

Electrical methods are used to map variations in electrical properties of the subsurface. The main physical property involved is electrical conductivity, which is a measure of how easily electrical current can pass through a material. Subsurface materials exhibit a very large range of electrical conductivity values. Fresh rock is generally a poor conductor of electricity, but a select group of metallic minerals containing iron, copper or nickel are very good conductors. Layers of graphite are also very good conductors.

The examples of good conductors mentioned above are quite rare. For most rocks, the electrical conductivity is governed to a large degree by the amount of water filling the pore spaces and the amount of salt dissolved in this water. Pure water has a very low electrical conductivity. On the other hand, seawater, which contains high levels of dissolved salts such as NaCl, is a relatively good conductor of electrical current. Groundwater can vary in salt content from fresh through brackish (slightly salty) to saline (similar in salt content to seawater) through to hyper-saline (more salty than seawater).

Electrical conductivity of rocks is not the only attribute which is of value to exploration geologists. A number of different electrical properties of rocks are measured and interpreted in mineral exploration. They depend on:

a) Natural currents in rocks – Self-potential method.

b) Polarizability of rocks – Induced polarization method.

c) Electrical conductivity or resistivity of rocks – Resistivity method.

d) Induction – Electromagnetic method.

Self Potential Method: Some materials tend to become natural batteries that generate natural electric currents whose effects can be measured. The self-potential method relies on the oxidation of the upper surface of metallic sulfide minerals by downward-percolating groundwater to become a natural battery; current flows through the ore body and back through the surrounding groundwater, which acts as the electrolyte. Measuring the natural voltage differences - usually 50-400 millivolts (mV), permits the detection of metallic sulfide bodies that lie above the water table. Other mineral deposits that can generate self-potentials are graphite, magnetite, anthracite, and pyritized rocks.

Induced Polarization: The passage of an electric current across an interface where conduction changes from ionic to electronic results in a charge buildup at the interface. This charge builds up shortly after current flow begins, and it takes a short time to decay after the current circuit is broken. Such an effect is measured in induced-polarization methods and is used to detect sulfide ore bodies.

Resistivity Method: Resistivity methods involve passing a current from a generator or other electric power source between a pair of current electrodes and measuring potential differences with another pair of electrodes. Various electrode configurations are used to determine the apparent resistivity from the voltage/current ratio. The resistivity of most rocks varies with porosity, the salinity of the interstitial fluid, and certain other factors. Rocks containing appreciable clay usually have low resistivity. The resistivity of rocks containing conducting minerals such as sulfide ores and graphitized or pyritized rocks depends on the connectivity of the minerals present. Resistivity methods also are used in engineering and groundwater surveys, because resistivity often changes markedly at soil/bedrock interfaces, at the water table, and at a fresh/saline water boundary.

Electromagnetic Methods: The passage of current in the general frequency range of 500-5,000 hertz (Hz) induces in the Earth electromagnetic waves of long wavelength, which have considerable penetration into the Earth's interior. The effective penetration can be changed by altering the frequency. Eddy currents are induced where conductors are present, and these currents generate an alternating magnetic field, which induces in a receiving coil a secondary voltage that is out of phase with the primary voltage. Electromagnetic methods involve measuring this out-of-phase component or other effects, which makes it possible to locate low-resistivity ore bodies wherein the eddy currents are generated.

A number of electrical methods described above are used in boreholes. The self-potential (SP) log indicates mainly clay (shale) content, because an electrochemical cell is established at the shale boundary when the salinity of the borehole (drilling) fluid differs from that of the water in the rock. Resistivity measurements are made by using several electrode configurations and also by induction. Borehole methods are used to identify the rocks penetrated by a borehole and to determine their properties, especially their porosity and the nature of their interstitial fluids.

Magnetic Methods

One of the most important tools in modern mineral exploration methods is magnetic survey. Magnetic surveys are fast, provide a great deal of information for the cost and can provide information about the distribution of rocks occurring under thin layers of sedimentary rocks - useful when trying to locate ore bodies.

When the Earth's magnetic field interacts with a magnetic mineral contained in a rock, the rock becomes magnetic. This is called induced magnetism. However, a rock may itself be magnetic if at least one of the minerals it is composed of is magnetic. The strength of a rock's magnetism is related not only to the amount of magnetic minerals it contains but also to the physical properties, such as grain size, of those minerals. The main magnetic mineral is magnetite (Fe_3O_4) - a common mineral found disseminated through most rocks in differing concentrations.

Measurements of the Earth's total magnetic field or of any of its various components can be made. The oldest magnetic prospecting instrument is the magnetic compass, which measures the field direction. Other instruments, which are appreciably more accurate include magnetic balances, fluxgate magnetometers, proton-precession and optical-pumping magnetometers.

Magnetic effects result primarily from the magnetization induced in susceptible rocks by the Earth's magnetic field. Most sedimentary rocks have very low susceptibility and thus are nearly transparent to magnetism. Accordingly, in petroleum exploration magnetic surveys are used negatively - magnetic anomalies indicate the absence of explorable sedimentary rocks. Magnetic surveys are used for mapping features in igneous and metamorphic rocks, possibly faults, dikes, or other features that are associated with mineral concentrations. Data are usually displayed in the form of a contour map of the magnetic field, but interpretation is often made on profiles.

It must be remembered that rocks cannot retain magnetism when the temperature is above the Curie point ($\approx$ 500°C for most magnetic materials), and this restricts magnetic rocks to the upper 40 kilometers of the Earth's interior.

When exploring for petroleum, magnetic surveys are usually made with magnetometers borne by aircraft flying in parallel lines spaced two to four kilometers apart at an elevation of about 500 meters. When searching for mineral deposits, the flight lines are spaced 0.5 to 1.0 kilometer apart at an elevation of roughly 200 meters above the ground. Ground surveys are conducted to follow up magnetic anomalies identified through aerial surveys. Such surveys may involve stations spaced only 50 meters apart.

A ground monitor is usually used to measure the natural fluctuations of the Earth's field over time so that corrections can be made. Surveying is generally suspended during periods of large magnetic fluctuation (magnetic storms).

Seismic Methods

Seismic methods are based on measurements of the time interval between initiation of a seismic (elastic) wave and its arrival at detectors. The seismic wave may be generated by an explosion, a dropped weight, a mechanical vibrator, a bubble of high-pressure air injected into water, or other sources. The seismic wave is detected by a Geophone on land or by a hydrophone in water. An electromagnetic Geophone generates a voltage when a seismic wave produces relative motion of a wire coil in the field of a magnet, whereas a ceramic hydrophone generates a voltage when deformed by passage of a seismic wave. Data are usually recorded on magnetic tape for subsequent processing and display. Seismic methods are of two kinds - Refraction methods and Reflection methods.

Seismic refraction methods: Seismic energy travels from source to detector by many paths. When near the source, the initial seismic energy generally travels by the shortest path, but as source to geophone distances become greater, seismic waves travelling by longer paths through rocks of higher seismic velocity may arrive earlier. Such waves are called head waves, and the refraction method involves their interpretation. From a plot of travel time as a function of source to geophone distance, the number, thicknesses, and velocities of rock layers present can be determined for simple situations. The assumptions usually made are that:

a) Each layer is homogeneous and isotropic (i.e., has the same velocity in all directions);

b) The boundaries (interfaces) between layers are nearly planar; and

c) Each successive layer has higher velocity than the one above.

The velocity values determined from time-distance plots depend also on the dip (slope) of interfaces, apparent velocities increasing when the geophones are updip from the source and decreasing when down dip. By measuring in both directions the dip and rock velocity, each can be determined. With sufficient measurements, relief on the interfaces separating the layers also can be ascertained.

High-velocity bodies of local extent can be located by fan shooting. Travel times are measured along different azimuths from a source, and an abnormally early arrival time indicates that a high-velocity body was encountered at that azimuth. This method has been used to detect salt domes, reefs, and intrusive bodies that are characterized by higher seismic velocity than the surrounding rock. Seismic waves may be used for various other purposes. They are employed, for example, to detect faults that may disrupt a coal seam or fractures that may allow water penetration into a tunnel.

Seismic reflection methods: Most seismic work utilizes reflection techniques. Sources and geophones are essentially the same as those used in refraction methods. The concept is similar to echo sounding - seismic waves are reflected at interfaces where rock properties change. The round-trip travel time, together with velocity information, gives the distance to the interface. The relief on the interface can be determined by mapping the reflection at many locations. For simple situations the velocity can be determined from the change in arrival time as source to geophone distance changes.

In practice, the seismic reflection method is much more complicated. Reflections from most of the many interfaces within the Earth are very weak and so do not stand out against background noise. The reflections from closely spaced interfaces interfere with each other. Reflections from interfaces with different dips, seismic waves that bounce repeatedly between interfaces ("multiples"), converted waves, and waves travelling by other modes interfere with desired reflections. Also, velocity irregularities bend seismic rays in ways that are sometimes complicated.

The objective of most seismic work is to map geologic structure by determining the arrival time of reflectors. Changes in the amplitude and wave shape, however, contain information about stratigraphic changes and occasionally hydrocarbon accumulations. In some cases, seismic patterns can be identified with depositional systems, unconformities, channels, and other features.

The seismic reflection method usually gives better resolution (i.e., makes it possible to see smaller features) than other methods, with the exception of measurements made in close proximity, as with borehole logs. In most exploration programs appreciably more money is spent on seismic reflection work than on all other geophysical methods combined.

Onshore Seismology

In practice, using seismology for exploring onshore areas involves artificially creating seismic waves, the reflection of which are then picked up by sensitive pieces of equipment called 'geophones' that are embedded in the ground. The data picked up by these geophones is then transmitted to a seismic recording truck, which records the data for further interpretation by geophysicists and petroleum reservoir engineers. The drawing shows the basic components of a seismic crew. The source of seismic waves (in this case an underground explosion) creates that reflect off the different layers of the Earth, to be picked up by geophones on the surface and relayed to a seismic recording truck to be interpreted and logged. Although the seismograph was originally developed to measure earthquakes, it was discovered that much the same sort of vibrations and seismic waves could be produced artificially and used to map underground geologic formations. In the early days of seismic exploration, seismic waves were created using dynamite. These carefully planned, small explosions created the requisite seismic waves, which were then picked up by the geophones, generating data to be interpreted by geophysicists, geologists, and petroleum engineers.

Recently, due to environmental concerns and improved technology, it is often no longer necessary to use explosive charges to generate the needed seismic waves. Instead, most seismic crews use non-explosive seismic technology to generate the required data. This non-explosive technology usually consists of a large heavy-wheeled or tracked-vehicle carrying special equipment designed to create a large impact or series of vibrations. These impacts or vibrations create seismic waves similar to those created by dynamite. In the seismic truck shown, the large piston in the middle is used to create vibrations on the surface of the earth, sending seismic waves that are used to generate useful data.

Offshore Seismology

The same sort of process is used in offshore seismic exploration. When exploring for natural gas that may exist thousands of feet below the seabed floor, which may itself be thousands of feet below sea level, a slightly different method of seismic exploration is used. Instead of trucks and geophones, a ship is used to pick up the seismic data and hydrophones are used to pick up seismic waves underwater. These hydrophones are towed behind the ship in various configurations depending on the needs of the geophysicist. Instead of using dynamite or impacts on the seabed floor, the seismic ship uses a large air gun, which releases bursts of compressed air under the water, creating seismic waves that can travel through the Earth's crust and generate the seismic reflections that are necessary.

2-D Seismic Interpretation

Two-dimensional seismic imaging refers to geophysicists using the data collected from seismic exploration activities to develop a cross-sectional picture of the underground rock formations. The geophysicist interprets the seismic data obtained from the field, taking the vibration recordings of the seismograph and using them to develop a conceptual model of the composition and thickness of the various layers of rock underground. This process is normally used to map underground formations, and to make estimates based on the geologic structures to determine where it is likely that deposits may exist.

Another technique using basic seismic data is known as 'direct detection.' In the mid-1970s, it was discovered that white bands, called 'bright spots', often appeared on seismic recording strips. These white bands could indicate deposits of hydrocarbons. The nature of porous rock that contains natural gas could often result in reflecting stronger seismic reflections than normal, water-filled rock. Therefore, in these circumstances, the actual natural gas reservoir could be detected directly from the seismic data. However, this does not hold universally. Many of these 'bright spots' do not contain hydrocarbons, and many deposits of hydrocarbons are not indicated by white strips on the seismic data. Therefore, although adding a new technique of locating petroleum and natural gas reservoirs, direct detection is not a completely reliable method.

Computer Assisted Exploration

One of the greatest innovations in the history of petroleum exploration is the use of computers to compile and assemble geologic data into a coherent 'map' of the underground. Use of this computer technology is referred to as 'CAEX', which is short for 'computer assisted exploration'.

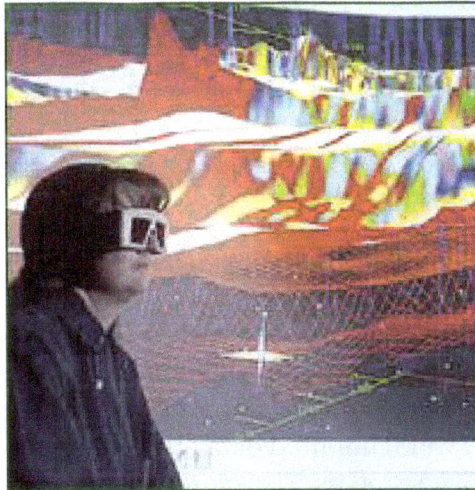

With the development of the microprocessor, it has become relatively easy to use computers to assemble seismic data that is collected from the field. This allows for the processing of very large amounts of data, increasing the reliability and informational content of the seismic model. There are three main types of computer-assisted exploration models: two-dimensional (2-D), three-dimensional (3-D), and most recently, four-dimensional (4-D). These imaging techniques, while relying mainly on seismic data acquired in the field, are becoming more and more sophisticated. Computer technology has advanced so far that it is now possible to incorporate the data obtained from different types of tests, such as logging, production information, and gravimetric testing, which can all be combined to create a 'visualization' of the underground formation. Thus geologists and geophysicists are able to combine all of their sources of data to compile one clear, complete image of subsurface geology. An example of this is shown where a geologist uses an interactive computer generated visualization of 3-D seismic data to explore the subsurface layers.

3-D Seismic Imaging

One of the biggest breakthroughs in computer-aided exploration was the development of three-dimensional (3-D) seismic imaging. Three-D imaging utilizes seismic field data to generate a three dimensional 'picture' of underground formations and geologic features. This, in essence, allows the geophysicist and geologist to see a clear picture of the composition of the Earth's crust in a particular area. This is tremendously useful in allowing for the exploration of petroleum and natural gas, as an actual image could be used to estimate the probability of formations existing in a particular area, and the characteristics of that potential formation. This technology has been extremely successful in raising the success rate of exploration efforts. In fact, using 3-D seismic has been estimated to increase the likelihood of successful reservoir location by 50 percent.

Although this technology is very useful, it is also very costly. 3-D seismic imaging can cost hundreds of thousands of dollars per square mile. The generation of 3-D images requires data to be collected from several thousand locations, as opposed to 2-D imaging, which only requires several hundred data points. As such, 3-D imaging is a much more involved and prolonged process. Therefore, it is usually used in conjunction with other exploration techniques. For example, a geophysicist may use traditional 2-D modeling and examination of geologic features to determine if there is a probability of the presence of natural gas. Once these basic techniques are used, 3-D seismic imaging may be used only in those areas that have a high probability of containing reservoirs.

In addition to broadly locating petroleum reservoirs, 3-D seismic imaging allows for the more accurate placement of wells to be drilled. This increases the productivity of successful wells, allowing for more petroleum and natural gas to be extracted from the ground. In fact, 3-D seismic can increase the recovery rates of productive wells to 40-50 percent, as opposed to 25-30 percent with traditional 2-D exploration techniques.

In addition to broadly locating petroleum reservoirs, 3-D seismic imaging allows for the more accurate placement of wells to be drilled. This increases the productivity of successful wells, allowing for more petroleum and natural gas to be extracted from the ground. In fact, 3-D seismic can increase the recovery rates of productive wells to 40 to 50 percent or greater, as opposed to 25 to 30 percent with traditional 2-D exploration techniques.

Three-D seismic imaging has become an extremely important tool in the search natural gas. By 1980, only 100 3-D seismic imaging tests had been performed.

Drilling Technique

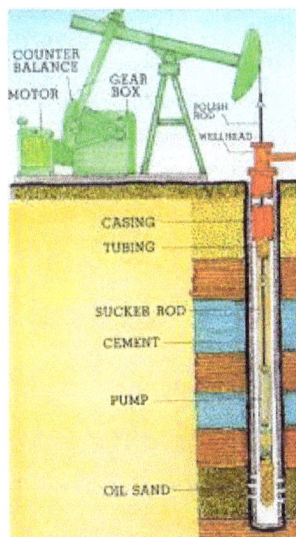

Drilling is the most important operation in the entire process of exploration and exploitation. It is used to recover the oil and gas deposit from the surface. Oil well drilling is totally different from the domestic well drilling because the oil wells are penetrated upto 5000m depth or even more where the formation pressure becomes so high. Drilling is of two types: -

Drilling Components

Various methods have been developed to drill different rock formation. Some of the important methods are:

1.   Cable tool method.

2.   Rotary drilling.

3.   Dyna drilling.

4.   Directional drilling.

5.   Offshore drilling.

Rotary drilling is generally used in onshore petroleum industry. All type of rotary drilling rigs usually has the same major components. These components are as following:-

(a) Derrick- To support the weight of the drilling equipment.

(b) Power Supply System- To provide power for running all machinery.

(c) Hoisting System- To raise or lower the drill string.

(d) Rotating System- To rotate the drill string.

(e) Circulating System- To circulate fluid down the pipe and throughout the hole.

(f) Well Control System- To maintain safety by controlling pressure imbalance in the well.

The following are the important assemblage of a drilling rig which is shown in figure.

(I)   Derrick Floor: - The steel frame tower which supports the tackle system for drilling, the drill pipe/ collar stands and running in and out of tools. The sizes of floor depend upon the capacity of rig.

(II)   Crown Block: - Sheaves mounted on beams and set on the top of the derrick used for hoisting and lowering of string equipments during drilling.

(III) Monkey Board: - The platform in the derrick on which the top man work during running in and out of drill strings.

(IV) Traveling Block: - Block with sheaves to which load is connected for hoisting or lowering a drilling operation. It is suspended from the crown block sheaves by steel wire rope as in block and tackle system.

(V)   Swivel: - A rotary joint under load through which mud is pumped under pressure and heavy rotating drill strings are suspended.

(VI)   Kelly: - A hollow square / hexagonal stem stacked to the top most drill pipe of the drill string which is rotated by the rotary table during drilling.

(VII) Draw Works: - A hoisted which is used for handling drill pipes, casing, tubing and other tools used in drilling. It is usually placed on the derrick floor and wire rope of the crown block is wound on its drum.

(VIII) Rotary Table: - A heavy geared circular steel having a hole at its center into which square or hexagonal hole bushing cab be fixed for engaging and rotating the drill string by Kelly. The table rotates in the horizontal plane and is normally driven by chains / carbon shaft from the draw works.

(IX)   Sub-Structure: - The steel frame work on which the derrick draw works and engines are installed.

(X)   Flow Nipple: - The Hollow Steel pipe through which the return of mud received in to the channels.

(XI)   Collar: - An excavator under the derrick designated to accommodate the well head fittings.

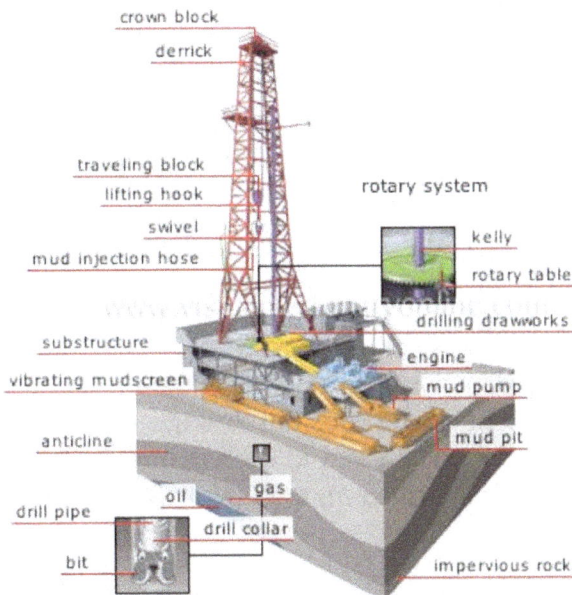

Figure: Drilling Rig and Its Components.

(XII) Blow Out Preventer (B.O.P): - An equipment to control the sudden outburst of gas, oil or water under high pressure encountered during drilling. It can either completely close the bore holes or close the annulus when drill string is inside the bore hole.

(XIII)   Cat Head: - A small spool retreated by the draw works on which a manila steel line is wounded and is used for making up or breaking off the driving drilling equipments on the derrick floor.

(XIV)   Shale Shaker: - A device on the rig to separate formation cuttings from the drilling and it comes out of the well.

(XV)   Brake Arm: - The lever which monitors the movement of the draw works driven.

Process

During rotary drilling three operations carried out simultaneously are i) a string of drilling pipe rotated with cutting bit; ii) the bit is lowered as the formation drills out and iii) the bit is cooled and lubricated.

The drilling rig is powered by 'diesel engines'. The essential piece of equipment of a rotary drilling rig is a derrick. The purpose of the derrick is to hold the travelling block assembly which holds the drill pipe. When a trip is made the derrick furnishes a most convenient place to stack the drill pipe vertically. When they are temporarily out of the hole, the depth capacity of drilling rig depend primarily on the size of draw works, which provide hoisting and rotating action, power from engine is transmitted to the draw works, then to the rotary table which rotate the kelly. The top of the kelly is attached to the swivel. Swivel allow the drill string to rotate while drill fluid passed through the crown block down to the draw works, drilling mud is passed through the bit with a jetting action. It assists the bit in cutting the hole. The fluid clean and cool the bit, cutting are carried by it through the annulus from which it passes over the shale shaker which separates the mud from cutting. The cutting goes into the pit and mud is recycled.

As the hole is drilled deeper additional length to the drill pipe is added by tool joints at derrick floor. The choice of the bit is an important factor to carry out drilling smoothly. Mostly diamond and tungsten carbide type of bit are being using in project area.

The rate of penetration depends on type of bit and bit rotation or pump pressure. The diameter of well is not kept always constant through out due to economic factors. The drilling is started with the 20" bit and in each casing diameter is reduced gradually. Finally 12.25" diameter hole drilled up to Cambay Shale and sometime up to Deccan Trap.

Drilling Fluid

The drilling fluid is normally called as mud. The mud is continuously pumped with high pressure into the well through drill pipes. It reaches the bottom of the well and comes out through the jets or nozzles of the bit; from there it takes the cuttings, makes its way (back) upwards through the annular space to the surface. Then it is passed into shale shaker to separate cuttings and other material. Shale Shaker or Vibrating Screen is built of two metal frames inclined at an angle of 12°-18° to the horizontal. A screen is fixed over each frame set on helical springs and resting on a firm footing. The screen is made of stainless wire. Generally 12-16 mesh screens are used. The frames are fixed with shaft. The vibration of screen destroys the thixotropic structure of the mud. Before going to shale shaker, the mud goes to gas separator, where the gas coming with mud is separated. Other items included in the fluid circulating systems are the de-sanders and de-silters. These are vortex cones which centrifugally removes undesirable materials that are fine enough to pass through the screen on the shale shaker.

After this, mud cleaner cleans the mud (remove clay size particle) and then passes it to the agitator or storage tank. In the storage tank barite powder, bentonite clay or mica etc. are added to mud to increase its density, viscosity etc. and then again goes to well and the process is continuously run during drilling.

In this system, liner is placed. The liner is comprised of two valves. Through one valve the mud is sent in to the well, while through the other valve, the mud comes out from the well. It works as a piston. At a time only one valve is opened. This is Mud Circulation System.

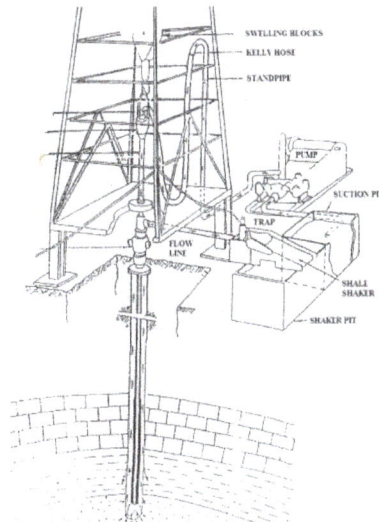

Figure: Mudflow System

Functions of Drilling Fluids

The important functions of drilling fluid (mud) are as following: -

(i) To clean the bottom of the hole, remove the cuttings and carry them to surface.

(ii) To cool and lubricate the drill bit and string.

(iii) To plaster the wall with mud cake and preventing the hole from caving.

(iv) To hold the cuttings and weight material in suspension while circulation is stopped.

(v) To support the weight of drill pipes and casing.

(vi) To transmit hydraulic horse power to the bit.

(vii) To acts as a medium of electrical well logging.

(viii) To ensure the maximum information about the formation penetrated.

(ix) Preventing corrosion of drill pipe.

Composition of Drilling Fluid

Main components of drilling fluid are as following: -

(a) Liquid Base: Water or oil.

(b) Colloidal Particles: It is used for a sedimentation stability of the system due to its transformation into a gel and also capable of plugging up pores and fine fractures in rocks.

(c) Weighting Material: To ensure pre assigned density of the fluid.

(d) Chemical Reagents: To regulate physico-mechanical chemical protecting adverse effect of environments. (e) Miscellaneous: Act as thinners, lost circulation materials, calcium removers, corrosion inhibitors, defoamers, emulsifiers etc.

Mud Logging

One of the most important problems encountered during drilling for hydrocarbons is the determination of fluid content of the porous formation opened by the bit, so that no oil or gas formations are overlooked. Geologging often referred to as mud logging in oil industry is a continuous monitoring system of the various parameters, like drilling rate, absence and type of gas in the mud system, mud loss/gain, examination of the drill cuttings, cores etc. for detecting presence or absence of hydrocarbons. The principal application of the geologging has been in drilling of wild cat or exploratory wells, where there is very little or no geological control. However the mud logging is helpful in gaining reliable qualitative information on the occurrence of hydrocarbons and this in connection with logging and interpretation derived from that, helps in giving more positive information about fluid contents of formations penetrated.

Principle

During drilling, mud is continuously pumped through the drill pipe to the bottom and out through the bit. It is then come back to the surface through annulus. While comes upto the surface it carries the cuttings and other drilled material. The mud is normally passed over a shale shaker where cuttings are separated. The chips are washed thoroughly and carried for further studies by the well site geologist. The studies are as follows: -

(I)   Evaluates the formation penetrated and the hydrocarbons shows encountered.

(II) Determination of subsurface contacts, lithological units for using the data on sub -surface mappings.

(III) Determination of physical characteristics of the sediments.

(IV) Interpretation of stratigraphy for using local and regional problems in sedimentology, structure, hydrocarbons accumulation.

Well Logging

Well logging operating are generally carried out in wells of all, either a wildcat well, exploratory, developing or producing well to know the various parameters of the well. Logging offers a way of gathering information's needed for both economic analysis and production planning. The geologist used to know the stratigraphy of the formation, the structural and sedimentary features and the mineralogy of the formation. The geo-physicist needs to know and to relate seismic references to specific horizons encountered in the drill section. The reservoir engineer needs to know the vertical and lateral extents of the reservoir, its porosity (type), permeability, the fluid content and its recoverability.

Logs provide either a direct measurement or indication of porosity both primary and secondary, permeability (to some extent), water saturation and hydrocarbon mobility, hydrocarbon type (oil and gas condensate), lithology, formation dip and structure, sedimentary environment, travel time of elastic wave in the formation.

Logging techniques in cased holes can provide much of the data which needs to monitor primary production and also to gauge the applicability of water flooding and monitor its progress when installed. In producing wells logging can provide measurement of flow rates, fluid type, pressure temperature, oil &/or gas saturation, points of fluid entry. Simply say logging when properly applied, can answer many questions ranging from basin geology to economics.

Well Completion

When drilling is completed, a well is prepared and filled with control equipments so that the well may safely produced oil/gas. This operation is known as well completion.

The well completion is the most important operational phase in the life of a well. This is the only way of communication with reservoirs. The effectiveness of communication with reservoir is main factor which control the reservoir drainage and economic life of well. The well completion involve the perforation of casing, well testing and activation of well through work over job.

Drill Stem Test

The drill stem test is used primarily to determine the fluids present in a particular formation and at the rate which they can be produced. The test is done in a bore hole filled with drilling mud. Pressure exerted by the drilling mud in the well prevents flowing out of fluid from the reservoir rock into the well. The drill stem has two expandable devices, called packers around it. The drilling stem is lowered into the well until one packer is just above the formation to be tested and other below. The packers are then expended to close the well above and below the formation. Sealing the well around the formation eliminates the pressure exerted by drilling mud on the formation. Water, gas or oil can flow out of the formation and in to the well. A trap door is opened on the drill stem and the formation fluids flow into and up the drill step.

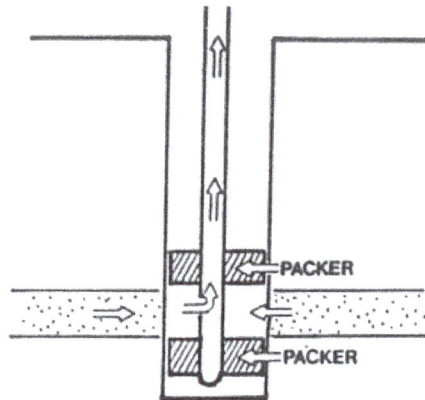

Figure: A Drill Stem Test.

Casing And Cementation

The first step in completing the hole is Casing. Casing is steel pipe that runs down the hole. Cement (slurry) is then pumped between the casing and sides of the well. Casing has three purposes. It stabilizes the well and prevents the sides from caving into the well. Casing protects fresh water aquifers that are often found near the surface. Casing seals off these reservoirs from pollution by mud during drilling and also petroleum during production. The casing also prevents the petroleum from being diluted by water from other formations during production. In completion of wells four types of casing are used. The size of casing and bit are as follows and as shown in figure.

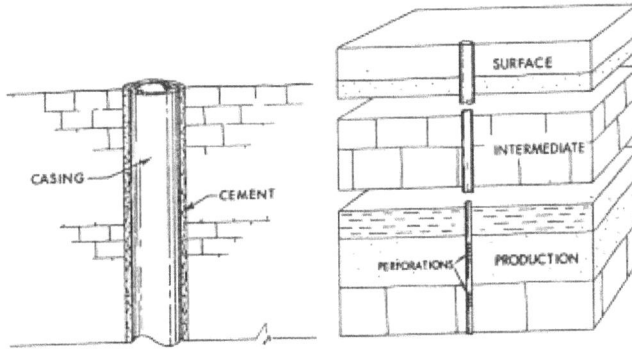

Figure: Casing in a well.
Figure: The casing program in a well, including strings of surface, intermediate, and production casing.

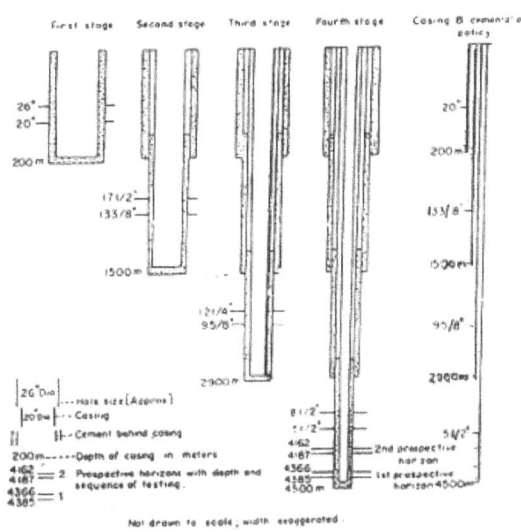

Figure: Telescopic casing system of oil-well.

(I)   Production casing- casing size = 5.5 inch; bit size = 8.5inch.

(II) Intermediate casing – casing size = 9.62 inch; bit size = 12.25inch.

(III) Conductor casing – casing size = 13.62 inch; bit size = 17.5 inch.

(IV) Surface casing – casing size = 18.62 inch; bit size = 26 inch.

Several items like guide shoe, float collar, centralizers and scratchers etc. are required. Guide shoe allows the movement of the casing tube down the hole and it is attached to the first joint of the casing. After running the joint halfway through the floor a centralizer is added (Figure A) and after running through full floor length another centralizer is added. Like this there may be 3 or 4 centralizers added to the drill string. Float collar is placed at a much higher height than the shoe and contains one way valve. The valve allows the casing to float down the hole so that the drilling fluid is prohibited entering the casing. After running the casing in the hole, filling fluid is essential to counter the buoyant effect and prevent damages caused by hydraulic pressure. Drill string often bounces while cutting hard rock. This causes damage to drill bit and drill string. To prevent such damages shock are kept on the drill bit. This absorbs all the shocks over this a drill collar is placed. Stabilizers are placed at 2or 3 places at a distance of 30 fit, centralizers and scratchers (Figure A & B) help in holding the casing away from the wall of the hole and abrades the mud when the casing tube is reciprocated. This procedure ensures proper distribution of cement around the pipe facilitating good bonding among pipe, cement and formation.

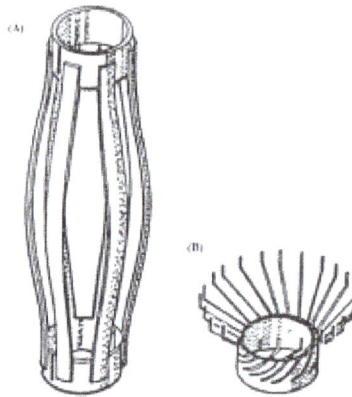

Figure: (A) Centraliser & (B) Scratcher

The cementing procedure also varies depending on the depth of the well. This informs the number of stages required for filling the annular space between casing and hole. The volume of cement to be pumped is calculated to ensure the filing of the annular. When the cement has set, the drill pipe can be run back and deepening of the hole can take place as usual.

Surface casing extends from the surface to 200 to 1500 feet deep depending on the area. It protects fresh water aquifers and prevents loose surface rock from caving into the well. Surface casing has a large diameter, for example, 13.62 inch outside diameter (OD). Intermediate casing is used to seal off problem formation such as salt or abnormal high pressure zones. It has a smaller diameter, for example, 8.62 inch OD. Production casing or oil string is run down to the producing formation. It has a smallest diameter, for example, 5.5 inch OD. The total length of a particular type of casing is called as string. A large diameter bit is used to drill the near surface portion of the well. After the surface casing string is run into the well, a small diameter bit is used to drill the next portion of the well. The well bore and casing become progressively are smaller with depth.

If the well bottoms out in the producing formation, a open hole completion can be used. The well is cased down to the top of the producing formation and left open below that. If the reservoir is loose sand, a screen, gravel pack or slotted liner is often put on the bottom of a well to prevent the well from becoming clogged. A cased well can also be completed with perforations (Figure A & B). Multiple completions on several reservoir rocks in a same well are done by perforations. The casing is run down the length of the well and a casing shoe closes off the bottom. A perforating gun is lowered into the well until adjacent to the reservoir rocks. The gun has either bullets or shaped explosive charges that are detonated to blow holes (perforations) in the casing, cement and reservoir rocks. This makes a very clean system; only fluids from the reservoir rock can flow into the well.

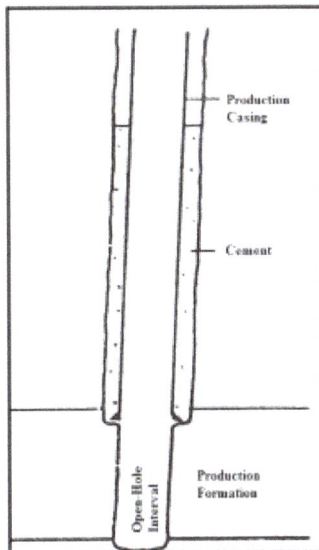

Figure: Open Hole Completion.

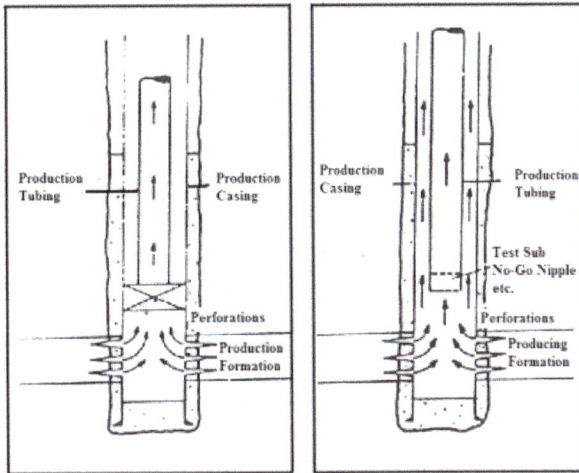

WITH PACKER   WITHOUT PACKER

Figure: (A) & (B) Cased Hole Completions.

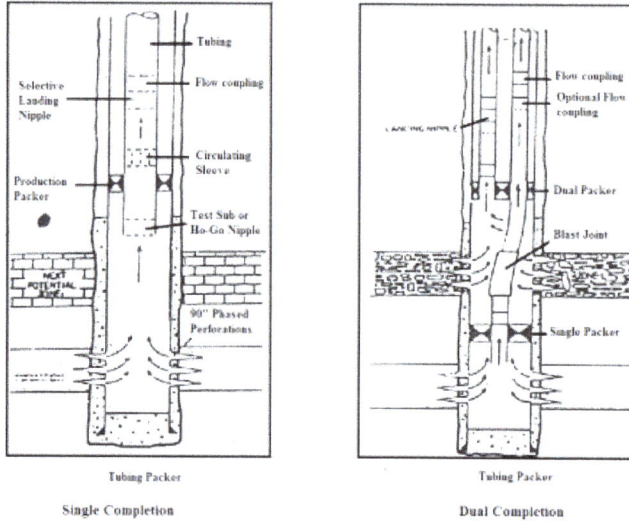

Figure: Multiple Completion

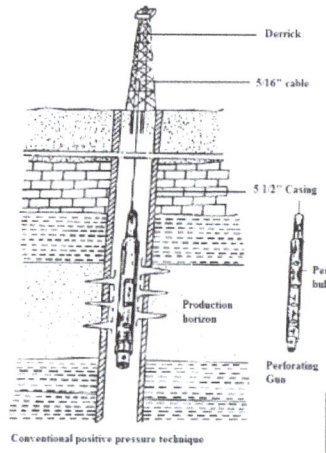

Figure: Perforating Operation.

Tubing

Small diameter pipe called tubing is run into the well to conduct the petroleum to the surface. Tubing ranges from 1.25 to 4.5 inches in diameter. It is suspended in the well from the surface down to just above the bottom. The pressure on oil in many reservoir rocks is enough to force the oil into the well but not to the surface. A pump is usually suspended in the well from the bottom of the tubing.

Surface Completion

In most oil wells, the oil has to be pumped to the surface through the tubing. The pumping unit is powered by an engine or prime mover that is usually electrical or diesel operated. The engine causes the walking beam to pivot up and down. As the end of the walking beam is raised and lowered a sucker rod is raised and lowered in the well. The sucker rod operates the pump at the bottom of the well. The oil or gas from the formation come and raises in tubing and goes to separator. Here oil and gas from the formation are separated out and oil is collected in tanks.

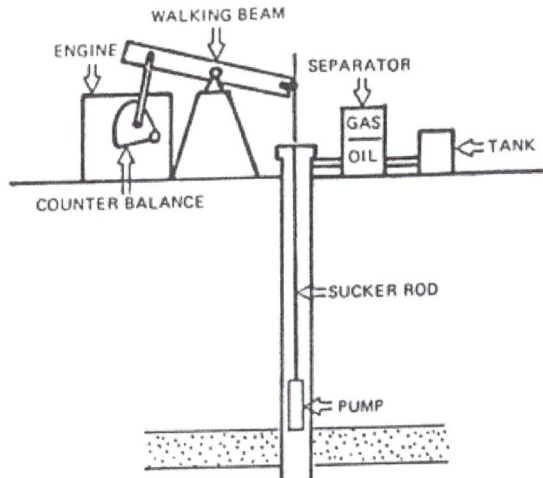

Figure: A Surface Pumping Unit.

Activation of Well

After a well is perforated, it is activated to include flow of formation fluid into the well. The activation of well can be done by following method such as:

(a) Swabbing

(b) Injection of compressed air or mixture of water and air down the annular.

(c) Injection of nitrogen gas at high pressure down to the annular.

Testing

After the well has been completed, a potential test can be run. This test will show the maximum gas and oil potential that the well can produce in a 24 hours period. A Productivity test is run to determine the effect of different production rates on pressures in the reservoir rocks. This is done by measuring the fluid pressure at the bottom of the well when it is not producing and then at several different rates of production. The test is used to calculate the maximum production rate of a well without damaging the reservoir.

Developing Of Field

It is well known that for the development of oil field, the drilling of development wells is necessary. For optimum number of wells, the following criteria has been used:-

(a) There should be minimum interference between the wells.

(b) The wells should be spaced in such a manner that they could be utilized in future for pressure maintenance or for enhanced oil recovery.

(c) The wells should be placed considering the isopay map, at the place where effective thickness is good and well may produce so much oil as to recover its cost etc.

Oil well spacing is often either 10, 20, 40 or 160 acres. Gas well spacing is often 160, 320 or 640 acres. A well spacing of 40 acres means that only one can be located on each forty acres of surface above the field.

Drilling Problem and Work Over

During the drilling, casing, well completion or when the well is flowing, several problems may arise which will be overcome by the work over. The following problems have been solved: -

If a tool is lost or the drill string breaks, the obstruction in the well is called as junk or fish. Special grabbing tools are used to retrieve the junk in a process, called fishing. In extreme cases explosives can be used to blow up the junk and then the pieces can be retrieved with a magnet.

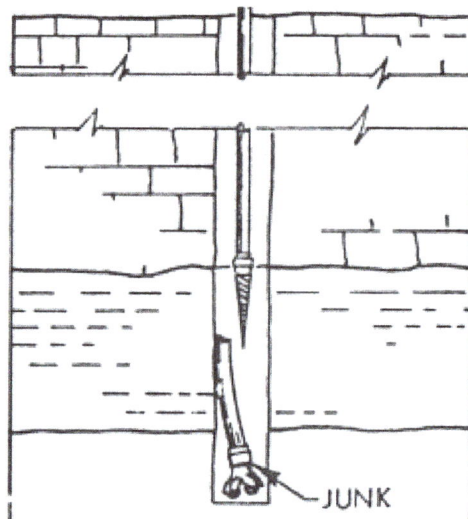

Figure: Fishing for Junk in the Well.

Some reservoir rocks can be damaged by forcing drilling mud into them. This can be caused by using too heavy an overbalance while drilling. The drilling mud clogs the pores or causes chemical or physical changes in rock. This decreases the rock's permeability near the well bore. Formation damage prevents or reduces production from the reservoir rock when the well is completed.

Lost circulation occurs when a very porous and permeable formation is encountered in the subsurface. The drilling mud flows into the formation without building up a filter cake. During lost circulation, more mud is being pumped down the well than is flowing back up. The level of mud in the mud pits falls and a float senses that there is a lost circulation problem.

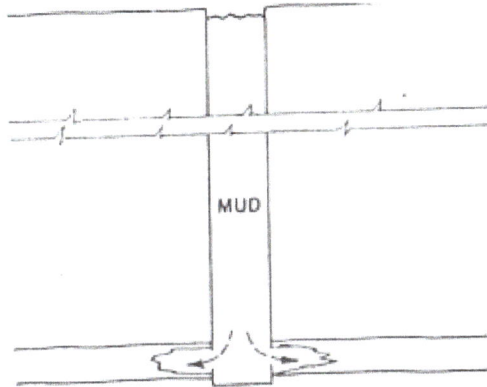

Figure: Lost Circulation caused by drilling mud flowing into a porous and permeable rock layer without building up a filter cake.

To solve these problems, fine grained fibrous materials such as mica flakes, ground pecan hulls, sugarcane hulls, shredded cellophane, and even shredded paper money pumped down into the well. The material got into the pores spaces of the lost circulation formation and swelled up, closing off the formation and solves the lost circulation problem.

An unexpected abnormally high temperature in the subsurface can cause a blow out. The overbalance is lost and the fluids flow out of the subsurface rocks into the well, is called as "kick". As the water, gas, or oil flows into the well, it mixes with the drilling mud, causing it to become even lighter and exert less pressure on the bottom of the well. The diluted drilling mud is called gas cut, salt-water cut or oil cut. The blows out preventers are immediately thrown to close the hole. The kick can be dangerous if it is caused by flammable natural gas or poisonous hydrogen sulphate gas. Sometimes the blowout occurs so fast that the drillers do not time to throw the blowout preventers and the result are disastrous. Slides and cables are located on the rig to evacuate the crew in such an emergency. If the blowout preventers are thrown in time, heavier drilling mud is pumped into the well through choke manifold to circulate the kick out. Abnormally high pressure is caused by the compaction of sediments.

A kick and possible blowout is detected by several different methods during drilling. As subsurface fluids enter the well during the kick, more fluids will be flowing out of the well than are circulating into the well. The sudden increase of fluid flow out of the well or rise of fluid level in the mud pit is detected by instruments. The drilling mud can also be continuously monitored for sudden changes in weight, temperature or electrical resistivity that would indicate the mud being cut by subsurface fluids. Another method is based on the principle that shale should become dense and less porous with depth as it is compacted. The density and porosity of shale can be determined from both well cuttings and well logs. If the shale density increases and porosity decreases are less than predicted from computations based on normal conditions, abnormally high pressure can be expected.

A blow out can also be caused by raising the drilling string out of the well. The drill string displaces a volume of drilling mud in the well. As the drill string is raised, the level of drilling mud falls in the well and the pressure is increased in the well, overbalance could be lost and a blow out could occur.

Drill String

Drill String is a string or a column of a drill pipe present on a rig which is used to transmit the torque and drilling fluids to the drill bit. The term is applied to the loosely assembled collection of the drill collars, drill pipe, drill bits and down hole tools. It is hollow inside hence it allows drilling fluids to be pumped down and circulated back up to the annulus. It is a combination of the bottom hole assembly, drill pipe and other tools that are used to make drill bits turn at the bottom of the borehole.

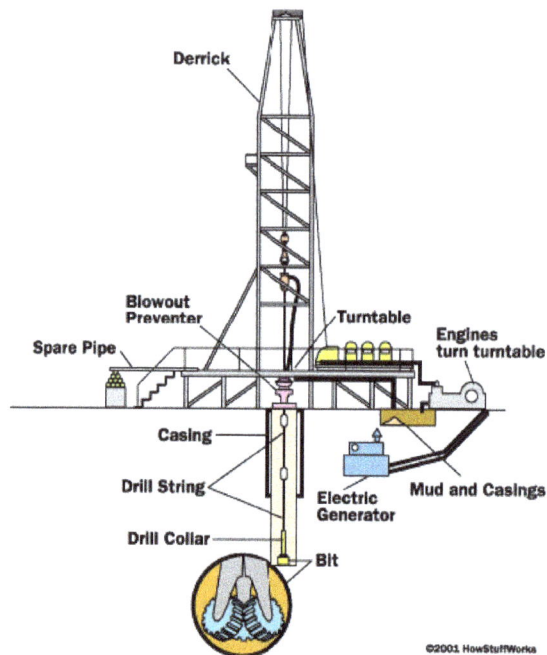

A Drill String is supported by a top drive that allows it to advance to the down hole and rotates it on the surface for driving the bits. It is used to maximize the hanging length and strength by using the lighter drill pipe with 5 inches of the outside diameter on the bottom for reducing its weight and by using stronger drill pipe with 5.5 inches outside diameter on the top where the bending stress is higher. The Drill String is run using a dual elevator system that uses two heavy lift elevators for supporting a pipe-tool joint instead of the kind of tooth that dies in slips.

Drill String Components

The drill string is typically made up of three sections:

- Bottom hole assembly (BHA)
- Transition pipe, which is often heavyweight drill pipe (HWDP)
- Drill pipe

Bottom Hole Assembly (BHA)

The BHA is made up of: a drill bit, which is used to break up the rock formations; drill collars, which are heavy, thick-walled tubes used to apply weight to the drill bit; and drilling stabilizers, which keep the assembly centered in the hole. The BHA may also contain other components such as a downhole motor and rotary steerable system, measurement while drilling (MWD), and logging while drilling (LWD) tools. The components are joined together using rugged threaded connections. Short "subs" are used to connect items with dissimilar threads.

Transition Pipe

Heavyweight drill pipe (HWDP) may be used to make the transition between the drill collars and drill pipe. The function of the HWDP is to provide a flexible transition between the drill collars and the drill pipe. This helps to reduce the number of fatigue failures seen directly above the BHA. A secondary use of HWDP is to add additional weight to the drill bit. HWDP is most often used as weight on bit in deviated wells. The HWDP may be directly above the collars in the angled section of the well, or the HWDP may be found before the kick off point in a shallower section of the well.

Drill Pipe

Drill pipe makes up the majority of the drill string back up to the surface. Each drill pipe comprises a long tubular section with a specified outside diameter (e.g. 3 1/2 inch, 4 inch, 5 inch, 5 1/2 inch, 5 7/8 inch, 6 5/8 inch). At each end of the drill pipe tubular, larger-diameter portions called the tool joints are located. One end of the drill pipe has a male ("pin") connection whilst the other has a female ("box") connection. The tool joint connections are threaded which allows for the mating of each drill pipe segment to the next segment.

Running a Drill String

Most components in a drill string are manufactured in 31 foot lengths (range 2) although they can also be manufactured in 46 foot lengths (range 3). Each 31 foot component is referred to as a joint. Typically 2, 3 or 4 joints are joined together to make a stand. Modern onshore rigs are capable of handling ~90 ft stands (often referred to as a triple).

Pulling the drill string out of or running the drill string into the hole is referred to as tripping. Drill pipe, HWDP and collars are typically racked back in stands in to the monkey board which is a component of the derrick if they are to be run back into the hole again after, say, changing the bit. The disconnect point ("break") is varied each subsequent round trip so that after three trips every connection has been broken apart and later made up again with fresh pipe dope applied.

Stuck Drill String

A stuck drill string can be caused by many situations:

- Packing-off due to cuttings settling back into the wellbore when circulation is stopped.

- Differentially when there is a large difference between formation pressure and wellbore pressure. The drill string is pushed against one side of the well bore. The force required to pull the string along the wellbore in this occurrence is a function of the total contact surface area, the pressure difference and the friction factor.

- Keyhole sticking occurs mechanically as a result of pulling up into doglegs when tripping.

- Adhesion due to not moving it for a significant amount of time.

Once the tubular member is stuck, there are many techniques used to extract the pipe. The tools and expertise are normally supplied by an oilfield service company. Two popular tools and techniques are the oilfield jar and the surface resonant vibrator. Below is a history of these tools along with how they operate.

History of Jars

The mechanical success of cable tool drilling has greatly depended on a device called jars, invented by a spring pole driller, William Morris, in the salt well days of the 1830s. Little is known about Morris except for his invention and that he listed Kanawha County (now in West Virginia) as his address. Morris received US 2243 for this unique tool in 1841 for artesian well drilling. Later, using jars, the cable tool system was able to efficiently meet the demands of drilling wells for oil.

8 inch drilling jar (red and white) on casings

The jars were improved over time, especially at the hands of the oil drillers, and reached the most useful and workable design by the 1870s, due to another US 78958 received in 1868 by Edward Guillod of Titusville, Pennsylvania, which addressed the use of steel on the jars' surfaces that were subject to the greatest wear. Many years later, in the 1930s, very strong steel alloy jars were made.

A set of jars consisted of two interlocking links which could telescope. In 1880 they had a play of about 13 inches such that the upper link could be lifted 13 inches before the lower link was engaged. This engagement occurred when the cross-heads came together. Today, there are two primary types, hydraulic and mechanical jars. While their respective designs are quite different, their operation is similar. Energy is stored in the drillstring and suddenly released by the jar when it fires. Jars can be designed to strike up, down, or both. In the case of jarring up above a stuck bottomhole assembly, the driller slowly pulls up on the drillstring but the BHA does not move. Since the top of the drillstring is moving up, this means that the drillstring itself is stretching and storing energy. When the jars reach their firing point, they suddenly allow one section of the jar to move axially relative to a second, being pulled up rapidly in much the same way that one end of a stretched spring moves when released. After a few inches of movement, this moving section slams into a steel shoulder, imparting an impact load.

In addition to the mechanical and hydraulic versions, jars are classified as drilling jars or fishing jars. The operation of the two types is similar, and both deliver approximately the same impact blow, but the drilling jar is built such that it can better withstand the rotary and vibrational loading associated with drilling. Jars are designed to be reset by simple string manipulation and are capable of repeated operation or firing before being recovered from the well. Jarring effectiveness is determined by how rapidly you can impact weight into the jars. When jarring without a compounder or accelerator you rely only on pipe stretch to lift the drill collars upwards after the jar releases to create the upwards impact in the jar. This accelerated upward movement will often be reduced by the friction of the working string along the sides of the well bore, reducing the speed of upwards movement of the drill collars which impact into the jar. At shallow depths jar impact is not achieved because of lack of pipe stretch in the working string.

When pipe stretch alone cannot provide enough energy to free a fish, compounders or accelerators are used. Compounders or accelerators are energized when you over pull on the working string and compress a compressible fluid through a few feet of stroke distance and at the same time activate the fishing jar. When the fishing jar releases the stored energy in the compounder/accelerator lifts the drill collars upwards at a high rate of speed creating a high impact in the jar.

System Dynamics of Jars

Jars rely on the principle of stretching a pipe to build elastic potential energy such that when the jar trips it relies on the masses of the drill pipe and collars to gain velocity and subsequently strike the anvil section of jar. This impact results in a force, or blow, which is converted into energy.

History of Surface Resonant Vibrators

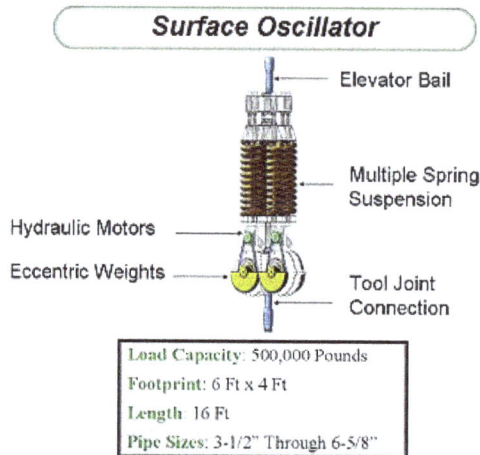

Oilfield Surface Resonant Vibrator

The concept of using vibration to free stuck objects from a wellbore originated in the 1940s, and probably stemmed from the 1930s use of vibration to drive piling in the Soviet Union. The early use of vibration for driving and extracting piles was confined to low frequency operation; that is, frequencies less than the fundamental resonant frequency of the system and consequently, although effective, the process was only an improvement on conventional hammer equipment. Early patents and teaching attempted to explain the process and mechanism involved, but lacked a certain degree of sophistication. In 1961, A. G. Bodine obtained US 2972380 that was to become the "mother patent" for oil field tubular extraction using sonic techniques. Mr. Bodine introduced the concept of resonant vibration that effectively eliminated the reactance portion of mechanical impedance, thus leading to the means of efficient sonic power transmission. Subsequently, Mr. Bodine obtained additional patents directed to more focused applications of the technology.

System Dynamics of Surface Resonant Vibrators

Surface Resonant Vibrators rely on the principle of counter rotating eccentric weights to impart a sinusoidal harmonic motion from the surface into the work string at the surface. Reference Three (above) provides a full explanation of this technology. The frequency of rotation, and hence vibration of the pipe string, is tuned to the resonant frequency of the system. The system is defined as the surface resonant vibrator, pipe string, fish and retaining media. The resultant forces imparted to the fish is based on the following logic:

- The delivery forces from the surface are a result of the static overpull force from the rig, plus the dynamic force component of the rotating eccentric weights.

- Depending on the static overpull force component, the resultant force at the fish can be either tension or compression due to the sinusoidal force wave component from the oscillator.

- Initially during startup of a vibrator, some force is necessary to lift and lower the entire load mass of the system. When the vibrator tunes to the resonant frequency of the system, the reactive load impedance cancels out to zero by virtue of the inductance reactance (mass of the system) equaling the compliance or stiffness reactance (elasticity of the tubular). The remaining impedance of the system, known as the resistive load impedance, is what is retaining the stuck pipe.

- During resonant vibration, a longitudinal sine wave travels down the pipe to the fish with an attendant pipe mass that is equal to a quarter wavelength of the resonant vibrating frequency.

- A phenomenon known as fluidization of soil grains takes place during resonant vibration whereby the granular material constraining the stuck pipe is transformed into a fluidic state that offers little resistance to movement of bodies through the media. In effect, it takes on the characteristics and properties of a liquid.

- During pipe vibration, Dilation and Contraction of the pipe body, known as Poisson's ratio, takes place such that when the stuck pipe is subjected to axial strain due to stretching, its diameter will contract. Similarly, when the length of pipe is compressed, its diameter will expand. Since a length of pipe undergoing vibration experiences alternate tensile and compressive forces as waves along its longitudinal axis (and therefore longitudinal strains), its diameter will expand and contract in unison with the applied tensile and compressive waves. This means that for alternate moments during a vibration cycle the pipe may actually be physically free of its bond.

Blowout

Blowout is nothing more than the uncontrolled expulsion of petroleum from a well. In the early stages of oil extraction, it is quite common for petroleum to be under pressure. This results from the fact that a fraction of nearly all petroleum reserves is gas. Because gases expand, a great deal of pressure can be created in petroleum reserves because they exist in confined and inflexible spaces. It is no somewhat like the pressure created by blowing air into a balloon. The more gas there is (air in the case of a balloon), the greater the pressure.

When a reserve is finally breached during drilling and there is an escape route for the pressure, hydrocarbon is carried up the bore hole and ejected onto the surface. Because the holes that are drilled to access a petroleum reserves are relatively narrow and because there can be a great deal of pressure in a well, hydrocarbon can be ejected up to 60 meters into the air. The more flammable hydrocarbons, like methane and propane, can easily be ignited as a result of friction during blowout, leading to fire and explosion. Underwater, the pressure can be enough to force oil out of a well even at depths of over a kilometer. The Deepwater Horizon blowout, which occurred in the Macondo Prospect oil field in the Gulf of Mexico, was the largest underwater blowout in history. Eleven crew were killed in the explosion. The well spilled an estimated 35,000 to 60,000 barrels of oil into the Gulf each day and was not repaired for over 3 months. Since that incident, blowout preventers have been under intense scrutiny. However, no better device has been proposed and standard blowout preventers are still being used. Many underwater drilling operations use more than one blowout preventer after the Deepwater Horizon accident.

Blowout Prevention

While not full proof, blowout preventers can help to stop blowout from occurring and have made drilling for petroleum safer and more environmentally friendly. Blowout preventers are often abbreviated as BOPs (pronounced B-O-P and not as 'bops' in the industry).

BOPs are mechanical devices and come in two basic types: ram and annular. Modern BOPs often use both mechanisms in tandem to help ensure the reliability of the system. In fact, most BOPs are constructed of at least one annular blowout mechanism atop several ram style mechanisms.

Ram Style Blowout Preventers

Ram blowout preventers were the first to be invented in 1922. Original ram BOPs were not intended to completely seal a well, but rather were intended to reduce the flow of oil to a manageable rate and allow time for capping. The ram BOP is based on the principle of a gate valve and uses two pairs of opposing steal plungers called rams, to restrict flow. The mechanism is simple, when pressure in the well moves from downward (into the well) to upward (out of the well) the rams slam shut like gates and slow or stop the flow of oil. Original rams, called pipe rams, simply sealed off the flow of petroleum around the drill pipe, but did nothing to prevent the flow through the pipe.

Modern ram BOPs use shear rams, which can cut through the drill pipe to completely stop the flow of oil. These ram BOPs rely upon hydraulic systems to provide the force needed to cut through strong metal drill pipe and drill string. To allow for recovery of expensive drill bits, most modern ram BOPs consist of both a shear ram and a pipe ram. The shear ram sits on top of the pipe ram and cuts the drill string to completely seal the well. Below that, the pipe ram is deployed at the same time to capture the drill string and prevent it from falling into the bore hole. This makes eventual recovery of the valuable bit feasible.

Annular Blowout Preventers

Annular BOPs were invented in 1952 and are often referred to as "Hydrils" after the company that produced them. Annular BOPs use rubber seals to close around the drill string and seal the well without cutting. This allows the drill string and bit to be removed from the well while pressure is maintained and blowout prevented. These BOPs work by forcing flexible rubber rings into tight wedges around the drill pipe and string.

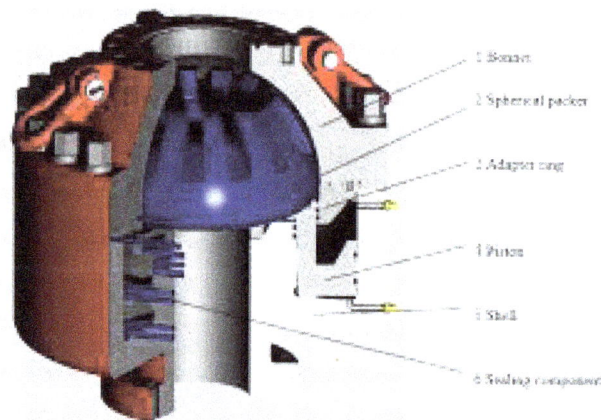

Annular BOPs are not as effective as ram BOPs in completely sealing a well. As such, annular BOPs are often stacked on top of ram BOPs to provide several levels of blow-out prevention. Annular BOPs necessarily rely upon hydraulic pressure to create and maintain their seals. In some cases, drilling can continue even when an annular BOP has been deployed.

References

* Geological-Mapping-Procedures-313799558: researchgate.net, Retrieved 16 March 2018

- Geochemical-Methods-of-Petroleum-Exploration: ijser.org, Retrieved 16 March 2018

- Exploration: naturalgas.org, Retrieved 25 June 2018

- Drill-string-1286: petropedia.com, Retrieved 14 May 2018

- Blowout-and-blowout-preventers: petroleum.co.uk, Retrieved 22 March 2018

Permissions

All chapters in this book are published with permission under the Creative Commons Attribution Share Alike License or equivalent. Every chapter published in this book has been scrutinized by our experts. Their significance has been extensively debated. The topics covered herein carry significant information for a comprehensive understanding. They may even be implemented as practical applications or may be referred to as a beginning point for further studies.

We would like to thank the editorial team for lending their expertise to make the book truly unique. They have played a crucial role in the development of this book. Without their invaluable contributions this book wouldn't have been possible. They have made vital efforts to compile up to date information on the varied aspects of this subject to make this book a valuable addition to the collection of many professionals and students.

This book was conceptualized with the vision of imparting up-to-date and integrated information in this field. To ensure the same, a matchless editorial board was set up. Every individual on the board went through rigorous rounds of assessment to prove their worth. After which they invested a large part of their time researching and compiling the most relevant data for our readers.

The editorial board has been involved in producing this book since its inception. They have spent rigorous hours researching and exploring the diverse topics which have resulted in the successful publishing of this book. They have passed on their knowledge of decades through this book. To expedite this challenging task, the publisher supported the team at every step. A small team of assistant editors was also appointed to further simplify the editing procedure and attain best results for the readers.

Apart from the editorial board, the designing team has also invested a significant amount of their time in understanding the subject and creating the most relevant covers. They scrutinized every image to scout for the most suitable representation of the subject and create an appropriate cover for the book.

The publishing team has been an ardent support to the editorial, designing and production team. Their endless efforts to recruit the best for this project, has resulted in the accomplishment of this book. They are a veteran in the field of academics and their pool of knowledge is as vast as their experience in printing. Their expertise and guidance has proved useful at every step. Their uncompromising quality standards have made this book an exceptional effort. Their encouragement from time to time has been an inspiration for everyone.

The publisher and the editorial board hope that this book will prove to be a valuable piece of knowledge for students, practitioners and scholars across the globe.

Index